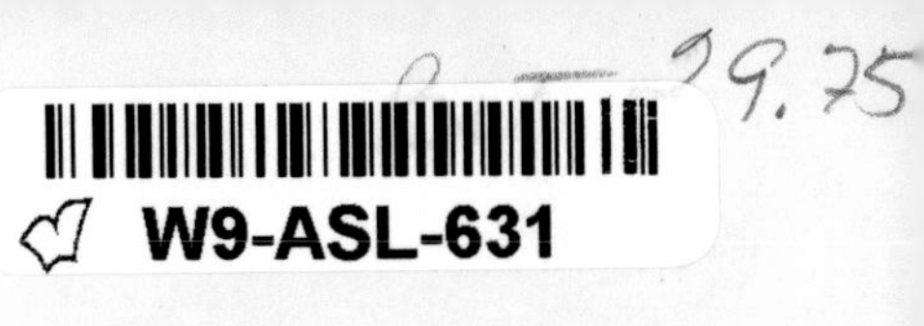

Misused Statistics

POPULAR STATISTICS

a series edited by

D. B. Owen
Department of Statistics
Southern Methodist University
Dallas, Texas

Nancy R. Mann
Biomathematics Department
UCLA
Los Angeles, California

1. How to Tell the Liars from the Statisticians, *Robert Hooke*
2. Educated Guessing: How to Cope in an Uncertain World, *Samuel Kotz and Donna F. Stroup*
3. The Statistical Exorcist: Dispelling Statistics Anxiety, *Myles Hollander and Frank Proschan*
4. The Fascination of Statistics, *edited by Richard J. Brook, Gregory C. Arnold, Thomas H. Hassard, and Robert M. Pringle*
5. Misused Statistics: Straight Talk for Twisted Numbers, *A. J. Jaffe and Herbert F. Spirer*

Other Volumes in Preparation

Misused Statistics

Straight Talk for Twisted Numbers

A. J. Jaffe
Senior Research Scholar, Retired
Columbia University
New York, New York

Herbert F. Spirer
Professor of Information Management
University of Connecticut
Stamford, Connecticut

MARCEL DEKKER, INC. New York and Basel

Library of Congress Cataloging-in-Publication Data

Jaffe, Abram J., [date]
Misused Statistics.

(Popular statistics ; 5)
Bibliography: p..
Includes index.
1. Social Sciences--Statistical methods.
2. Statistics. I. Spirer, Herbert F. II. Title.
III. Series.
HA29.J29 1987 300'.1'5195 86-16237
ISBN 0-8247-7631-3

MARCEL DEKKER, INC.
270 Madison Avenue, New York, New York 10016

Current printing (last digit)
10 9 8 7 6 5 4 3 2 1

PRINTED IN THE UNITED STATES OF AMERICA

Dedicated to two great teachers
who knew a misuse when they saw one
and knew how to set it right

William F. Ogburn
President
American Statistical Association, 1931
Editor
Journal of the American Statistical Association, 1920–1925

Eugene D. Homer
Formerly of New York University

Preface

Every day we see **misuses of statistics** which affect the outcomes of elections, change public policy, win arguments, get readers for newspapers, impress readers, support prejudices, inflame hatreds, provoke fears, and sell products.

It doesn't have to be that way! Statistics—the science of using, summarizing, and analyzing numbers—*can be* and often is correctly used and when it is, we are all better off.

We, the authors, are by nature measurers and questioners, which is why we are in this profession. And because we questioned, we started a series of articles on misuses of statistics which appears in the *New York Statistician.* It has attracted great interest, including that of Maurits Dekker, who suggested we expand the series into a book.

Our goal is to help you to be a critical observer of the statistical scene. We want to help you in asking the right questions by showing you dozens of examples of misuses of statistics, and providing some of the questions you can and must ask to *see things as they are.* As you will discover in reading this book, you are never done with asking questions about statistics. And each time you ask a question you get more information.

To help you, we have classified misuses into arbitrary categories and grouped them into chapters. You will see that misuses come in clusters; sometimes so many misuses are in one analysis that we must call it a "megamisuse." You may think a classification other than ours is better. If you feel that way, good! Because this means that you are also becoming a questioner in this search after a special kind of truth.

In some of our examples, we show not only a misuse but also a proper use of statistics. However, our book is not a textbook or report of investigations, which is where proper uses are and should be described. Don't let the presence of so many misuses in one book discourage you from using statistics! There have been many instances where the proper use of statistics has been of great benefit. We hope that the number of such uses will be increased by your awareness of the issues we raise here.

Have we ourselves misused statistics? We hope not, but as you will see, it is hard to avoid an inadvertent misuse and we are as fallible as any other statistician.

We have prepared this book for people whose statistical knowledge is less than that of the professional statistician, but it is meant to serve professionals as well. Because this book illustrates statistics as applied to real-world problems, it should be invaluable as an adjunct to statistical textbooks, both graduate and undergraduate.

A. J. Jaffe
Herbert F. Spirer

Acknowledgments

We are indebted to Maurits Dekker for suggesting that our many examples of misuses of statistics which appeared in the *New York Statistician* could, and should, be made into a book; to Marilynn Deuker of the Department of Statistics at the University of Connecticut for critical evaluations and much advice; to Louise Spirer who served as a nonstatistical reader and whose keen eye for logical and editorial errors was invaluable; to Rosalind Fink, the Affirmative Action Officer of Columbia University, who advised us on affirmative action (Chapter 12) and comparable worth (Chapter 13); to Carolyn Sperber of Columbia University, who also helped with comparable worth; to Professor Robert Lewis of the Department of Geography of Columbia University who helped us with information about the state of statistics in the USSR; to Edward Spar who provided us with information on the condition of the statistical apparatus in the United States (Chapter 14); to the contributors of examples of misuses to the *New York Statistician*; to Will Lissner, editor of the *American Journal of Economics and Sociology*, who encouraged us and published our preliminary paper on misuses; and to our two universities (and their excellent libraries) for their support of our work.

We sincerely thank these people and any who may have been inadvertently omitted. But we alone are responsible for the contents of this volume.

Contents

Misused Statistics

1
Introduction

Truth will ultimately prevail where there is pains [sic] taken to bring it to light.

George Washington

When you can measure what you are speaking about and express it in numbers you know something about it; but when you cannot measure it, when you cannot express it in numbers, your knowledge is of a meager and unsatisfactory kind.

Lord Kelvin

Misuse of Statistics: *Using numbers in such a manner that—either by intent, or through ignorance or carelessness—the conclusions are unjustified or incorrect.*

The Authors' Definition

Numbers—all numbers—are important. Even a zero can be important; the only difference between a check for $100 and one for $1000 is a single zero. Let's explore this idea further.

I. The Importance of Numbers

Statistics—numbers—have been used by humans for thousands of years, perhaps even before modern *Homo sapiens sapiens* appeared on the earth. We know of censuses taken 3000 years ago and numbers are recorded on the most ancient tablets and stelae dating back to the beginning of the use of written (or inscribed) symbols.

Prehistoric humans must have used quantitative thinking to decide whether the game slain or the amount of gathered fruits, vegetables, and nuts was adequate to feed the group. Even if our ancestors did not use our sophisticated methods of counting, their very survival suggests that they could estimate as we do. They must have compared the number of people and the amount of food available. At the kill of a mastodon, most of the members of a small group must have been able to see that there was more meat available than the group could eat. On the other hand, if the hunter brought back only one rabbit, he or she must have known that it was not enough for more than one or two people. Our prehistoric ancestors must have used quantitative reasoning in the same ways as today's cooks do in making similar decisions.

Today, we need sophisticated (i.e., statistical) methods for dealing with numbers; our concerns go beyond hunting and gathering. For example, one statistic of great importance to people in the United States is the number of people living in each state. The U.S. Constitution mandates that the members of the House of Representatives be apportioned among the states in accordance with each state's population. Thus the U.S. Bureau of the Census counts the population every ten years. At present, the bureau must use complex methods because of the large size of the population and the complexity of the task.

In this day and age you cannot escape numbers, nor can you escape sophisticated statistical methods. Nor do most people want to. Imagine baseball without batting averages, number of bases stolen, runs scored, games won and lost, and pitching records. Or football without the number of yards gained, number of passes completed, lengths of passes, and so forth. People in sports who do not see themselves as quantitative thinkers continually make decisions based on statistics: Who is the best available pinch hitter in a given situation, measuring "best" in terms of batting averages under similar circumstances? What is the ranking from highest to lowest of the tennis players entered in the U.S. Open tournament? The players are ranked by a complicated formula.

If your interest is business and finance rather than sports, think of the importance of stock market indexes, commodity prices, stock and bond prices, corporate profits, bond ratings, and so forth. Decisions to buy or sell are made on the basis of the information conveyed by these statistics. Are you concerned with marketing? If so, you will need numbers and statistical methods to answer key questions for making decisions: How many potential customers are there for a particular product or in a particular geographic region? How rich or how poor are the potential and actual customers? What are the characteristics of our potential customers?

It is easy to show how statistics—numbers—are crucial to our political survival. Numbers are the basis for decisions as to what actions to take and which policies to accept, reject, or formulate. Government officials constantly make decisions based on numbers. For example, the unemployment rate is watched by elected officials, administrators, and the public when deciding whether the government should take action to influence the economy: Shall we start a job training program for the unemployed in a particular region or industry? Shall the term of payment of unemployment benefits be increased or decreased?

Skeptics may argue that political decisions are made on another basis than the relevant numbers, but they should remember that political decisions almost invariably take into account the number of voters with a given opinion. Rarely does an elected official vote against an issue which is supported by a large proportion of his constituents.

Some facts of life regarding the use of statistics:

The misuse of statistics can lead to wrong decisions. This is the heart and purpose of our book: to show how to recognize misuses of statistics that lead to wrong decisions.

It is an unpardonable sin to use numbers for propaganda purposes. The deliberate misuse of statistics to "prove" a political point has often led to policies which caused serious harm to peoples, organizations, and countries.

Numbers need only be as correct as is necessary for the purpose at hand. For example, during the 1985 drought in the northeastern United States, depletion of reservoirs was a cause of concern. Is it essential to have an exact measure of the status of the reservoirs to take appropriate action? Whether the reservoirs are 45%, 50%, or 60% full makes no difference. If the reservoir status is in that range, the responsible officials know that they are "too empty" and, with a forecast of no rain, those officials know that action must be taken.

Most statistics are not exact and may not be exact enough for the purposes at hand. But how inexact? Sometimes we need highly accurate statistics; if the statistics you have are exact enough, you can make a decision, otherwise you can't. For example, in the apportionment of representatives to the states, the loss or gain of a congressman can occur for a difference of 1000 people in a state with a population in the millions. In relative terms, an error of 1000 out of 10,000,000 amounts to an error of one part in ten thousand. This looks like an extremely small error, and it would be in most circumstances. But here it looms large because of its possible consequences.

II. Why Knowledge of Misuses Is Important

A. Learning from Misuses

We can all profit from the study of misuses and misinterpretations of statistics. The general public, elected and appointed governmental officials, business people, journalists, TV and radio commentators, medical doctors, dentists, and students can learn how to detect a misuse of statistics. Even professional statisticians and teachers can profit, for none of us knows all that there is to know.

For most people—the nonstatisticians—this book's goal is to show how to spot misuses in the media (including scholarly and professional journals) and in public statements. The key question is: What shall I believe? This question is best answered when the reader or listener can evaluate statistical arguments.

For professional statisticians (including teachers) and those who must interpret and publish articles which include or are based upon statistics, the goal is to avoid statistical misuse. Report writers too often fool themselves, as well as the public, by misusing statistics through ignorance or neglect. Even skilled professionals can make errors, as will be seen when we discuss the remarkable case of Samuel George Morton.

We want to show students and nonstatisticians how to evaluate statistical inputs in their decision making. You do not need a knowledge of statistical theory to deal with many of the misuses which affect decision making in public and private life. While it is common to sneer at the subject and say that "you can make statistics say anything," it is only through *misuse* that you make statistics say "anything." Good statistics tell only the truth!

We also hope to show students of statistics that they can learn from the study of misuses. This is an approach that they will not find in most elementary statistics textbooks. As Friedman and Friedman tell it:

> Most introductory courses in statistics focus on how to use statistics rather than how to *avoid misusing* statistics. Many textbooks offer a "cookbook" approach to the analysis of data, without advising the reader what will happen to the recipe if one ingredient is left out. . . .
>
> It is our belief that one can learn more from studying misapplications of statistics than from perfect applications. Many disciplines now use case studies as teaching aids. These case studies not only demonstrate the correct way of solving problems, but also show the effects of incorrect decisions, enabling students to learn from others' mistakes. It is one thing to teach students about the assumptions underlying various statistical procedures; it is quite another matter to show realistic examples of the consequences of violating these assumptions. [1, pp. 1–2]

We do not propose *not* teaching students the right way to apply statistical tools. However, while not all students of elementary statistics will become statisticians, virtually all students will use and apply statistics in their future professional or private lives. As the statistical theories of the classroom fade away with time, the lessons learned from the study of misuses will become more important.

In this book, we describe some of the many misuses which we have observed and explain why they are misuses. When you have a full understanding of how others create misuses, you are better equipped to avoid creating misuses yourself and to recognize them when they appear in newspapers, in politicians' speeches, in advertisers' claims, and even in scholarly journals. Most of the cases cited are real but we have included several hypothetical cases of misuses committed by Dr. K. Nowall, an imaginary researcher. These hypothetical cases are typical of misuses which have occurred but we have simplified the situations to give our readers more accessible illustrations.

We also describe some examples of the proper use of statistics. This is not done to take on a function of books on statistical methods but to illuminate our examples of misuses. As the inadequacies of a poorly-designed engine can be seen more easily by comparing it to a well-designed one, a misuse of statistics can be more easily understood by comparing it to a proper use.

B. What Is a Misuse of Statistics?

The four misuses which follow are the ones we consider most important:

1. Lack of knowledge of the subject matter
2. Faulty, misleading, or imprecise interpretation of the data and results
3. Incorrect or flawed basic data
4. Incorrect or inadequate analytical methodology

The presence of any one of these flaws is enough to invalidate the results and create a misuse. In addition, it is not unusual for a number of misuses to occur in a given case; we call this special horror case a *megamisuse*.

If the investigator knows the subject and interprets the data and results correctly, the last two flaws (*data, analytical methods*) can still cause a misuse. Table 1.1 shows how these two factors work together to yield misuse or proper use.

Correct and incorrect are absolute terms, and in evaluating methods and data we frequently must assign relative measures of correctness. For example, you might think that getting data for the number of accidents during a particular

Table 1.1 How the Two Factors of *Data* and *Analytical Methods* Lead to Misuse

Analytical methods	Data	
	Correct	Incorrect
Correct	Proper use	Misuse
Incorrect	Misuse	Misuse

time period on a particular stretch of highway would be a simple counting process. But it can be quite difficult to count accidents in such situations as we discuss in Chapters 4 and 5. Despite your best efforts, you may be in the unpleasant situation of not being able to say whether the counts you have are correct or incorrect. All you may be able to say is that, *within a particular context*, the data are "reasonably correct," or "probably incorrect," or some similar relative statement.

Why? Because errors can be made in the physical counting process, reports can be lost, dates or locations of occurrences can be incorrect, different observers may use different definitions of accidents, and so forth. If the only known errors are of this kind and you have reason to believe they are small (for your purposes), then these data are "reasonably correct." However, if you are evaluating the effect of an accident prevention program by comparing one month to another and you expect only small changes, then (in this context) these data are "probably incorrect."

You may have a case in which several different methods of statistical analysis can be used on the data. Unfortunately it is not always possible to have one uniquely correct procedure for a set of data. Furthermore, different outcomes may result from the use of different analytical methods. If more than one conclusion can be drawn, then you have a misuse of statistics unless you show and explain all of the possible results. If you show only one conclusion or interpretation, ignoring the alternative procedure or procedures, then you have committed a misuse.

Sometimes you will find it hard to make a clear distinction between the proper and improper use of statistics. Political polling is a good example. The interviewers can ask respondents how they feel about candidate x and many respondents will say that they like candidate x. Then the interviewers ask whether the respondent will vote for a particular candidate. A large proportion (e.g., 36%) say that they "do not know." Of those who say they will vote for a particular candidate, x gets the largest proportion. But what is that proportion? Let us say the results are:

Candidate x: 40%

Candidate y: 24%

Don't Know: 36%

What can you do with these results? You could conclude that Candidate x "leads the pack." Conversely, you could conclude that the majority (the Don't Knows plus those who favor Candidate y) do not favor x but for reasons known only to themselves 36% refuse to publicly acknowledge their intent to vote for y.

Which interpretation is correct? The data and analytical procedures of this survey may be correct, but the results are inconclusive: only the election gives the answer. We have shown two of the many possible interpretations of these results, neither of which can be shown to be superior to the other in the absence of additional information. If only one interpretation is reported, then it is a misuse of statistics; if both or more are reported, then we have a proper use of statistics.

III. Florence Nightingale Used Statistics Properly to Save Lives

The statistical work of Florence Nightingale, founder of modern medical nursing, is an example of the *proper* use of statistics. Nightingale was appalled by the lack of proper medical and nursing care and the unsanitary conditions she found in the army hospitals during the Crimean War. Her remarkable study led to a series of successful medical and nursing programs which greatly reduced human suffering.

She collected data which showed decisively and quantitatively that more soldiers died from diseases attributable to unsanitary conditions than from battle wounds. Excluding men killed in action, the mortality rate "peaked at an annual rate of 1174 per 1000. . . . This means that if mortality had persisted for a full year at the rate that applied in January, and if the dead soldiers had not been replaced, disease alone would have wiped out the entire British army in the Crimea" [2, p. 133].

Her remarkably effective lobbying, supported by novel graphic presentations of the mortality statistics, brought an immediate response from the British public and government. She was given support to make sanitary reforms, causing the death rate to drop dramatically. This she also documented superbly. Her statistical analyses were the key factor in motivating the modernization of the British armed forces medical programs and, subsequently, the introduction of a new standard of nursing and hospital care:

> Nightingale . . . strongly influenced the commission's work, [The Royal Commission on the Health of the Army] both because some of its members were her friends . . . and because she provided it with much of its information. . . . She wrote and had privately printed an 800-page book [*Notes on Matters Affecting the Health, Efficiency and Hospital Administration of the British Army*] . . . which included a section of statistics accompanied by diagrams. Farr [a professional statistician] called it "the best [thing] that ever was written" either on statistical "Diagrams or on the Army." . . . Nightingale was a true pioneer in the graphical representation of statistics . . . much of Nightingale's work found its way into the statistical charts and diagrams [Farr] prepared for the final report. . . . Nightingale had the statistical section of the report printed as a pamphlet and widely distributed. . . . She even had a few copies of the diagrams framed for presentation to officials in the War Office and in the Army Medical Department. [2, pp. 133, 136]

IV. Some Consequences of Misuses

The results of the misuse of statistical analysis can be slow to surface. In electrical engineering, by contrast, an incorrectly designed circuit can result in an immediate and verifiable failure. But in setting public policy, a statistical misuse can lead to wrong decisions with consequences that may not be obvious for decades. By that time, more data will have accumulated—but someone has suffered in the interim. To some extent, technological misuses of statistics are self-correcting and do not often appear in print. For that reason, many of our examples of misuses are drawn from nontechnical situations.

A. An Unintelligent Misuse to Support Racism

Racism is a persistent problem, and an early misuse of statistics in its support has persisted from the early 20th century to the 1980s. An early argument in this cause said that the genetically unintelligent quit school young. The attendant development of the Intelligence Quotient (IQ) seemed to justify this argument. Jay Gould recently studied this belief:

> Of all invalid notions in the long history of eugenics—the attempt to ''improve'' human qualities by selective breeding—no argument strikes me as more silly or self-serving than the attempt to infer people's intrinsic, genetically based ''intelligence'' from the number of years they attended school. . . . The genetic argument was quite popular from the origin of IQ testing early in our century until the mid-20s, but I can find scarcely any reference to it thereafter. [3]

Gould shows that subsequent statistical analyses reveal little, if any, relationship between years of schooling and intelligence. More years of schooling do not necessarily indicate more intelligence.

B. Be Fruitful and Multiply

But favorite statistical misuses do not necessarily die, nor even fade away. In 1983, Lee Kwan Yew, Singapore's Prime Minister, resurrected this same misuse of statistics as the basis for statements that better educated Singaporans should have more children. Lee stated that the 1980 census reported that women with no education had an average of three-and-a-half children, or more than double the number borne by women with a university degree. Lee argued that since women with university degrees are ''more intelligent,'' they should have more children. He then said that 80% of intellectual achievement resulted from nature and only 20% from environment, including schooling. Where did the Prime Minister get his data? ''In my reading [says Gould], the literature on estimates of heritability for IQ is a confusing mess—with values from 80% . . . all the way down to Leon Kamin's contention . . . that existing information is not incompatible with a true heritability of flat zero [as paraphrased in 3, p. 27]. How many generations does it take for a known statistical misuse to pass away?

V. Deliberate or Inadvertent?

Are we surprised when misuses and errors support the viewpoint of the originator? Most people seem to accept this as normal. President Reagan overestimated the growth in work time needed to pay taxes by a factor of almost eight and the growth in the percentage of earnings taken by the Federal government in taxes by more than ten times [4]. Similarly, Samuel George Morton, who had the world's largest collection of pre-Darwinian skulls, manipulated the data to support his conclusions on the inferiority of non-Caucasians [5].

A. Is There a Smoking Gun?

The analyst must be careful in assigning motives, for we are all prone to make errors in our favor. It is tempting to say both of the above examples are deliberate misuses of statistics and to attribute conspiratorial intent to the originators. However, without having the smoking gun—explicit statements of the intent to deceive—you can't know a person's motives.

President Reagan's incorrect statistics may come from "number illiteracy" rather than deliberate misstatement or an intent to mislead. As an economist noted, "For Mr. Reagan, it is not just a question of numbers, *which are not his strong point* . . . [italics ours] [6]." Morton published all his raw data; thus, we can show his manipulations of the data and derive the correct results. Because he did not attempt to hide the data, we may conclude that he did not intend to mislead his readers and probably did not realize that he was manipulating data to fit his preconceived ideas.

B. No Literary License Is Granted to Misusers of Statistics

We have seen a case in which the originator of a misuse admits that it is deliberate. In their article in *Science*, Pollack and Weiss stated that "the cost of a telephone call has decreased by 12,000%" since the formation of the Communication Satellite Corporation [7]. Letters to the editor of *Science* pointed out this misuse of statistics. Pollack replied: "Referring to cost as having 'decreased by 12,000%,' *we took literary license to dramatize the cost reduction* [italics ours] [8]." If you want to avoid statistical misuses, don't confuse fictional literature with the presentation of statistical findings.

C. Play Fair When You Play Statistics

You have the right to suspect a deliberate intention to mislead when someone refuses to supply the raw data and/or sources for the data. Professional statisticians rarely refuse to give sources and raw data. If the report supplies sources of data or the data itself, then you cannot reasonably suspect deliberate deception.

> Gross flouting of procedure and conscious fraud may often be detected, but unconscious finagling by sincere seekers of objectivity may be refractory. . . . I propose no cure for the problem of finagling: indeed, I . . . argue that it is not a disease. The only palliations I know are vigilance and scrutiny. [5]

Be vigilant and scrutinize!

2
Categories of Misuse

Our little systems have their day;
They have their day and cease to be.

Tennyson

I must create my own System or be enslav'd by another Man's.

Blake

I. Introduction

For many years the *New York Statistician*, the official journal of the New York Area Chapter of the American Statistical Association, published articles and notes on the misuse of statistics. In 1975, as the editors, we incorporated this material into a series entitled ''Misuses of Statistics.'' Journal readers continued to send in examples which, together with the large number we collected elsewhere (over 200 in all), are the basis of this book.

A. Categories

We have organized statistical misuses into five major categories, which we hope are adequate and mutually exclusive. In addition, because similar offenses often occur within categories, we have set up subcategories to simplify identification. Many examples of misuses involve several offenses; we call these multiple misuses *megamisuses*. There is no limit as to the kinds of abuses and new ones are being committed daily; thus our categories are arbitrary and subject to change with increased knowledge.

There are two common denominators to misuses: First, the writer or investigator does not understand the basic data which are used or reported. Second, the writer or researcher does not understand the basic quantitative methodology. The methodology may be simple or complex; without understanding of how it was devised or when particular techniques are appropriate, misuse results.

The five categories of statistical misuses and their subcategories are:

1. Lack of knowledge of subject matter
2. Quality of the basic data
   a. Flaws
   b. Bias
   c. Bad definitions
3. Preparation of the study and the report
   a. Design of the study and report
   b. Graphic presentation
   c. Interpretation of findings
   d. Presentation of results
4. Statistical methodology
   a. General
   b. Erroneous calculations
   c. Regression
   d. Sampling
   e. Variability
   f. Verifiability
5. Deliberate suppression of data

The rest of this chapter contains brief discussions of each of these categories to set the stage for the balance of the book. In Chapters 3 through 10, we present more detailed examples of each type of misuse, explain why we consider them to be misuses, identify some of the people who originate misuses, describe some of the statistical methods which are misused, and show the dangers of misuse. In Chapter 11, we discuss the principle of Ockham's Razor, which bears directly on many types of misuses. In Chapters 12 through 14, we analyze particular instances of misuses in affirmative action, the creation of mythical numbers which we call "ectoplastistics," and the actions of past and present governments of diverse political philosophies.

II. Discussion of Misuses

A. Lack of Knowledge of Subject Matter

Sometimes researchers or writers do not understand the basic data with which they are working, do not know how to compute a statistic, how to formulate a hypothesis, how to get existing information, or how to test the results for validity. Thus we have a biologist suggesting that the U.S. Census start using samples to get information instead of taking a complete census [1]!

The Constitution mandates a complete count (or as complete as can be obtained) for the purpose of allocating Congressional representatives among the states. But (as we discuss later) sample data, such as the number of children born to a woman, have been collected for several decades for purposes which are not based on a *legal* requirement for a census.

B. Quality of the Basic Data

1. Flaws

Sometimes the basic data are not available or cannot be *validated*. Statistical analyses affecting public policy may be based on data obtained in situations where the facts cannot be known or verified. Some recently-published examples of such data are the number of abortions performed in a third world country, the number of deaths in a remote area due to starvation, the number of people living in jungles, and the number of illegal aliens living in the United States. All are typical of such data.

In other cases, the facts are known but are *misrepresented* and the person doing the analysis is unaware of this misrepresentation.

Data may be invalid for the purpose at hand. The ratio of U.S. taxes paid to the total profits of a multinational corporation (which includes overseas earnings) is not a valid measure of the ratio of taxes to profits, since taxes may have been paid in foreign countries [2].

The data may not be *comparable*, as when comparing voter turnout percentages by states. Some states "purge" the registration lists to eliminate nonvoters (based on the reasonable presumption that nonvoters may have left the state or died); other states have no purge rules at all, and there are intrastate variations in purging rules [3].

Some data are *invalid* under any circumstances because of measurement error. Measurement errors plague the polling analyst as much as the engineer and physical scientist. The errors can arise in ways that are discussed later.

2. Bias

Data may be biased because of the manner of collection. Leading questions in questionnaires can be as subtle as a sledge hammer. A "referendum" on nutrition asks who opposes the marketing of junk foods, with no option for not opposing and no definition of "junk foods" [4].

In a professional questionnaire designed to detect bias, respondents can oppose a constitutional amendment to prohibit abortion but also support a constitutional amendment to protect the life of the unborn child [5]. In this case, a question designed to determine an underlying attitude produces a different response depending on the wording of the question. Do you think these different answers represent different attitudes?

3. *Bad Definitions*

Countries that report economic statistics often use different definitions. Japan's Finance Ministry, the U.S. Department of Commerce, and the International Monetary Fund use three different definitions for looking at trade surpluses between the United States and Japan. One definition may indicate a surplus while another indicates no surplus—an important difference when Congress is considering protectionist legislation [6].

C. Preparation of the Study and the Report

1. *Design*

If an experiment is to be valid, it must be designed to give clear and identifiable results. To introduce the important and complex subject of the design of experiments, we will examine a real-life experiment.

Almost all of us have or have had wisdom teeth, so we base our example of experimental design on a research paper which was concerned with testing a new method for reducing infection in the holes ("sockets") left when wisdom teeth are removed [7, as discussed in 8]. In this example, the reduction in infection was to be measured by comparing the new method to an existing treatment. But individuals vary greatly in their potential for infection and their response to medication depends on their age, health, inherent characteristics of the immune system, and so forth.

How are we to separate out these "confounding" factors? One way would be to simultaneously extract a tooth from each of a pair of identical twins. But it is unlikely that both would need an extraction at the same time and to perform extractions for no other reason than to complete the experiment is unethical. However, in a reasonable time period we might be able to find two groups of individuals needing extractions who are similar in as many of the confounding factors as possible. The group receiving the new treatment is the *experimental group* and the group getting the existing treatment is the *control group*. (We might even be able to find individuals who need two extractions at one time.) The control set is composed of the sockets (after extraction of the tooth) in each mouth that get the existing treatment and the experimental set is composed of the tooth sockets receiving the new treatment.

How are we to assign the treatment to the different teeth to assure that there is no systematic effect? There *might* be a difference in response among many humans between right and left sides of the mouth. We do not say there is, but experiments with other mammalian characteristics have shown differences between sides. To reduce the effect of this factor, we "randomize" the assignment of treatment to side of the mouth, using some random process (tossing dice, random number tables, random assignment by computer, etc.).

How do we make sure that the dental surgeons apply the new and existing treatments so they do not influence the results, consciously or unconsciously? Supervision is one approach, but in general it is better to make a "blind" experiment in which the surgeon does not know which treatment he is applying, the new or the old.

Alas, determining the intensity of an infection is largely subjective, a determination made by the surgeon with support from the patient and some objective measurements. How are we to reduce any conscious or unconscious influence on the results in this case? Once again, we can use a blind approach, with both the surgeon making the evaluation and the patient unaware of which tooth has been exposed to which treatment. When both the application of the trial method and the evaluation are performed blind, we call it a *double blind* experiment.

This example is a brief, and simplified, introduction to some of the pitfalls that can occur in the design of experiments. If you are a professional researcher, it is expected that you know the basic principles (and more!) and that you will use them when you design experiments.

Despite the fact that professionals (and even many nonprofessionals) know these principles, papers of greater and lesser importance that violate one or more of these principles still appear in refereed scientific journals. Sometimes the conclusions published in scientific papers are reported in the media and, unfortunately, these media reports seldom include information on the design of the study so that the concerned reader can make an informed judgment. To their credit, some media do, on occasion, succinctly communicate those design facts which are important to the nonprofessional reader or listener, as the *New York Times* does when reporting its survey results.

2. *Graphic Presentation*

Graphic misrepresentation is a frequent misuse in presentations to the nonprofessional. The granddaddy of all graphical offenses is omitting the zero on the vertical axis (as shown in Figure 6.1). As a consequence, the plot is often interpreted as if its bottom axis were zero, even though it may be far removed. This can lead to attention-getting headlines about "dramatic increases." A modest, and possibly insignificant, change is amplified into a disastrous or inspiring trend. How are we to "put a magnifying glass" to changes which may be significant? By clearly indicating that the vertical axis does not go continuously to zero or by focussing on the changes, as we discuss in Chapter 6.

3. *Interpretation of the Findings*

How many are the ways to misinterpret findings! How easy it is to create an exciting headline by projecting to the population as a whole results from a sample drawn from some unique part of the population. Many researchers also teach and when they wish to test a hypothesis, they find that their students are a ready source of subjects. The researcher-teacher administers a survey to a small group of students (usually the ones in his classes, a special segment of a special segment) and if the researcher wisely doesn't project the results to the population of the United States as a whole, some headline writer will, as we show in Chapter 8.

Periodically there is a great deal of public concern for education in the United States. High school student performance on the Scholastic Achievement Test (SAT) is often taken as a measure of the quality of education. In 1974, a school introduced a competency program and then showed a gain of 31% in the SAT pass rate over the pass rate of 1971. The school's administrators claimed that this increase was a result of the program. But the administrators did *not* report that the rate had already risen considerably in 1973, one year before the program was introduced [9].

4. *Presentation of Results*

Results are one thing, the way they are presented is another, especially if essential elements are omitted. Results are often presented as being significant without any indication of the sample size (from which the sampling error can be deduced). Union Carbide does not show "Don't Knows" in an advertisement reporting on a professionally-conducted poll of attitudes on energy, an unfortunate omission [10]. When the "Don't Knows" are placed in the table describing the results, the reader's conclusion is likely to be different than the one proposed in the advertisement.

Simply repeating numbers *ad infinitum* can result in misinterpretation of the facts, as Malcolm W. Browne has described [11]. Even if the numbers are correct, the impression which is desired may be false: "A lot of people are afraid of numbers, so numbers make wonderful cudgels for winning arguments."

Browne describes a television commercial in which the salesperson tells the audience that his product is superior because it contains 850 milligrams of "pain reliever" while other products have only 650 milligrams. Questions which arise are: Is the "pain reliever" the same in both products? If not, which pain reliever is best for a particular case? If they are the same, is this pain reliever efficacious? Is it bad to take more than 650 milligrams in one dose? By not dealing with such questions but simply repeating "850 milligrams," the presentation misuses numbers—statistics—to create an impression which may be incorrect.

D. Statistical Methodology

1. *General*

There are many different statistical methods. Elementary textbooks list dozens, and new ones are constantly being developed and reported. If a researcher uses the wrong method to analyze a specific set of data, then the results may be incorrect. This is a misuse. If a researcher uses the "wrong" typical value, information may be hidden or disguised and a wrong result may be drawn. For example, there are several methods for determining a typical value: arithmetic mean, median, mode, geometric mean, harmonic mean, mid-extreme, and so forth. Each has a specific computational formula and each is suited to particular data and interpretation needs. To determine the "typical" salary in a community where a few individuals have extremely high salaries, the informed analyst will use the median, as more representative of the salaries earned by people in the community, rather than the arithmetic mean (which can create a false impression, as we show in Chapter 7).

2. Erroneous Calculations

Neither careful proofreading of newspaper copy nor the referee process for academic journals guarantees error-free computation. We see many examples of incorrect results in computation involving only elementary arithmetic and have published, alas, such errors ourselves. In a *Wall Street Journal* article on the effect of heart surgery, we found four incorrect results from faulty addition. None of these errors affect the conclusions, but (surprise!)—the errors emphasize the desired effect [12]. We hope that some of these errors can be avoided now that the personal computer has replaced the abacus, but the analyst or reporter must still know which totals should reconcile. It's easy enough to do: a rough total using a calculator can serve as a check.

The use of computers does not always mean a reduction in errors. It may even make things worse—a lot worse. While the computer may make fewer arithmetic errors, it can create new kinds of errors which may be even harder to find.

A team of researchers in the field of aging said that increasing numbers of the elderly moved to metropolitan areas in the last decade. They used data on computer tapes supplied by the U.S. Census Bureau to arrive at this conclusion. But there was an error in the data which:

> . . . involved a change in the way the bureau arranged the information on the computer tapes of its 1980 data . . . In previous censuses, . . . those who moved to the United States from abroad were not counted as part of interstate migration within America.
>
> In 1980, . . . the census counted 173,000 people who came to the United States from abroad during the last decade and "lumped them into the group" of elderly people moving from nonmetropolitan areas. . . . Thus those who had moved from foreign countries were counted as part of interstate migration. . . . The researchers failed to notice that the two kinds of migration had been lumped together. [13]

Does the Social Security system contribute to a reduction in savings, and hence, a reduction in investment capital? A computer programming error resulting in erroneous calculations led to a multibillion-dollar overestimate of the negative effect of Social Security on national saving [14]. All that computer power was brought to bear and it produced a false result. Interestingly, the director of the project, while acknowledging the error, stated that he has not changed his beliefs.

3. Regression

Simply stated, *regression* is the process of drawing a line through a series of points in a two-dimensional plot, as shown in Figure 2.1. One of the things you can do when you have drawn a *regression line* is predict the value of *Y* (along the vertical axis) by knowing the value of *X* (along the horizontal axis). (See Figure 2.1.)

Some misuses of regression are easily explained. For example, you are right to suspect any prediction of a value of *Y* based on values on *X* that are outside the range of the data from which the line was obtained (A, B in Figure 2.1). But if you don't see the plot of the points, as is often the case, then you may not know that you are using a value of *X* outside the range of observed values.

Linear regression, the form usually used, gives you the formula for a

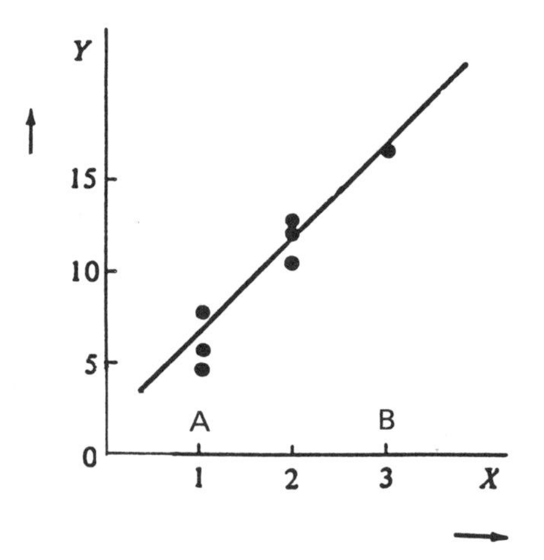

Figure 2.1 An example of a *regression line*. The black dots are the observed data points having particular values of the variables on the *X*- and *Y*-axes; the line is obtained by "fitting" it to the observed data points. A common use of the regression line is to predict values of *Y* based on knowing a value of *X*. When the value of *X* lies outside the range of observed points (as indicated by "A" and "B"), you cannot be sure the relationship holds.

straight line and some statistics on how well the line "fits" the data. If you rely on such summary measures and do not see the plot of the points, then you can fail to see important relationships. In Figure 2.2, we show two plots of points with identical regression lines. As you can see, the sets of points are quite different, but the equations you would get from a calculator or a computer printout are identical!

4. Sampling

Redbook surveyed its readership on sexual attitudes using a questionnaire inserted in the magazine. Of the more than 3,000,000 subscribers, 20,000

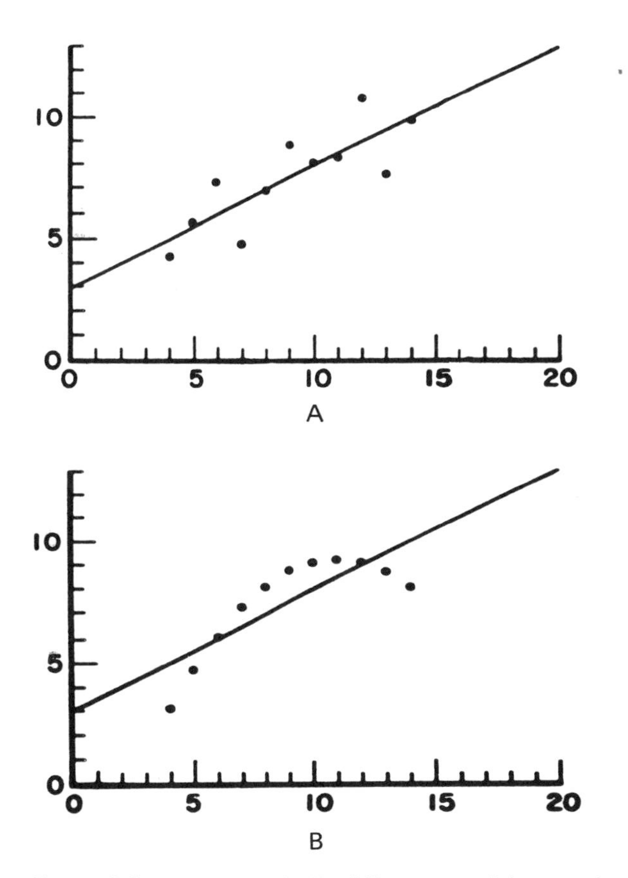

Figure 2.2 Two dramatically different sets of data are shown with the fitted regression lines. The lines are identical! Imagine what a misuse of statistics it would be to use the line in (A) to project a value on the Y-axis for a value of 20 on the X-axis. (*Source*: F. Anscombe, Graphs in statistical analysis, *The American Statistician*, *27*(1):17 (1973). Used by permission. All rights reserved.)

women and 6000 men selected themselves and responded. Why did they select themselves? Were the reasons such that these respondents form a special group? Would you project the responses from this self-chosen group to the entire population of the United States? Although the *Redbook* article was exemplary in cautioning readers that the sampling (interviewing only part of the whole population) was not random and could not be projected to the whole population of subscribers, the *New York Post* article describing the results gave no cautions about projecting the results to the general population [15].

Charges of sexual abuse in New York child care centers were the impetus for attempts to estimate the number of employees at the centers who had criminal records. Mayor Koch released results of an initial sample of 82 employees in which about 44% of the employees checked reportedly had criminal convictions. The headlines were dramatic. Subsequently George Gross, the city's Human Resources Commissioner, reported on an analysis of a sample of 6742 fingerprints of employees and showed that 253, or about 4%, were found to have criminal records.

The officials involved were primarily concerned about sample size:

> The Mayor released those first figures (from the sample of 82 employees)—despite the objections of some mayoral advisors who were concerned that the sampling was too small to be valid. . . .
>
> "With this kind of huge sample, my feeling is that the numbers will probably hold up . . ." Commissioner (Gross) said. [16]

The real issue is the one they do not deal with: how the samples were selected. Were the first 82 sample employees chosen randomly from the population of about 20,000 such employees in the city? If they were, then it is reasonable to project that between 33% and 55% of the employees had criminal records. If the true proportion of employees with criminal records is below 5%, then it is extremely unlikely—almost impossible—to have obtained the results cited by the Mayor, *if the sample was random*. And if it was not random, then we can make no projections to the population of all child care center employees. Thus, we have no basis for an estimate of what proportion of the employees had criminal records.

On the other hand, Commissioner Gross gives no information as to how his sample of 6742 was chosen. We can draw no meaningful conclusions about either estimate from the results as stated, because we do not know how these samples were chosen.

The use of polls to project the results of elections is quite common today (as we discuss in detail in Chapters 9 and 10). Common sense dictates that if you want to project the outcome of an election—that is, the voting behavior of those who will vote in the election—you should get a sample consisting of people eligible to vote in the election. And yet we find:

> . . . an ABC/*Daily News* poll this month showed the Mayor [Koch, of New York] with sixty-seven percent of the vote—forty-eight points ahead of Miss Bellamy and sixty-one points ahead of Farrell [Mayor Koch's challengers in the upcoming primary election]—although the fact that two-thirds of those polled were not registered Democrats and thus couldn't vote in the primary cast some doubt on its precision. [17, p. 79]

Need we say more?

5. Variability

When survey results are given as proportions and without mention of the sample size, there is no way to estimate the sampling error—that error in the results which is due to the chance effects of sampling. Are you surprised that the advertisement does not tell the sample size when it states that "9 out of 10 MERIT smokers (are) not considering other brands" [18]? You have good cause to be concerned when two medical researchers give several percentages concerning vasectomy and cancer of the cervix without giving the sample size [19].

Has the jobless rate risen when the headline says: "Jobless rate in U.S. up slightly . . . to 5.7% from 5.6%" [20]? You need a knowledge of the sampling error (variability) to answer this question. In this case, although the headline was misleading, the author of the article knew enough about the subject to state in the article that the jobless rate had held "essentially constant." But if an article reporting a statistical result is reluctant to make the comparison between the change and the observed variability, then the reporter should give some measure so that the readers can make their own evaluation. In Chapter 9 we show you that you can easily evaluate a survey when you have been given enough information about sample size and the method of selection. A few papers, including the *New York Times*, routinely include, as a part of every survey, both the necessary information and the results of the calculations.

6. Verifiability

Statistical results often are held in such low esteem that it is imperative that the results be verified. Unfortunately, we often cannot confirm the results obtained by authors. The reasons are many: computations on the data given do not yield the same results as those reported; insufficient data are given so that readers cannot check the results (as in the case of a widely-copied statement that 10,000 to 20,000 people die each year of cancer-causing food additives [21]); there is insufficient information on how the sample was chosen; the sources of data are not stated; or the author has collected and processed the data so that it is impractical, if not impossible, for anyone else to check the reported results.

Sometimes, the use of a computer may muddle, rather than clarify, statistical results. A group of physicists, computer scientists, and others complained that some computer programs were too complicated and thus so difficult to read that a reader could not determine exactly what was done to the data in order to verify the results. This group is now reluctant to accept the results given by researchers who make extensive use of this type of computer analysis [22].

7. Deliberate Suppression of Data

As we discussed in Chapter 1, it is hard to know what is in someone's mind when that person takes an action. What appears to be a deliberate action may in fact be an error or an act which the research investigator sincerely believes to be correct and appropriate. On the other hand, we often see cases in which governmental and organizational officials suppress data and information and clearly state their intentions. In Chapter 14, we show how the United Nations responded to political pressure and stopped distribution of statistics on Taiwan. We also show cases in which nations selectively stop the collection and publication of national statistics.

III. Summary

Has the misuse and abuse of statistics declined with the increase in the proportion of the population with higher education? Has the advent of inexpensive high-powered computers reduced the number of misuses? We don't know the answers to these questions but we do know that results of complex computer analysis have as much potential for a misuse as a misleading graph. Educators, the educated, and professionals need a basis for dealing with misuses of statistics. We feel that this is best learned through a study of the misuses themselves rather than with more theoretical information.

Methods of mass communication (''the media'') have gained in speed and access. We have the instant access of television reporting with its (to some degree understandable) lack of depth in coverage of complex topics; the interaction between the observed and the observer in cable TV ''surveys''; and phenomena such as the effect of statistical projections on the behavior of voters whose behavior is being projected.

Sample surveys and polls have become a cultural institution and can have great influence. Often, the full spectrum of misuses described in this book is incorporated into published poll results. The consequence is poor decision making by a misled public and policy makers. These abuses of statistics occur frequently in the media and public policy statements, despite warnings from those who produce the polls and surveys. The public, the policy makers, and those who report on the surveys and polls appearing in the media must understand the limitations of statistical methods and how statistics can be misused. Through public discussion of misuses and abuses of statistics, we hope to improve the use of statistical methods. Vigilance and scrutiny are the hallmarks of the informed researcher, reporter, and reader.

3
Know the Subject Matter

Not well understood, as good not known.

Milton

Knowledge is of two kinds; we know a subject ourselves, or we know where we can find information upon it.

Samuel Johnson

I. Introduction

When a writer or researcher does a quantitative analysis of a topic, he or she must be knowledgeable about the subject matter. The writer or researcher may apply all the appropriate "schoolbook" statistical tools to a real world problem—the methodology may be correct, the arithmetic flawless—but without a knowledge of the subject matter, the results probably will be incorrect and flawed: a misuse of statistics.

There are many cases where there are conflicting sets of statistical data. Some are good, some are bad, and some may be useful, but only if the researcher knows enough about the subject to take the good and reject the bad. The writer or researcher needs knowledge of the subject matter to make sensible use of the sources and data.

II. Dr. K. Nowall, A Fictitious Example

A. The Unexpected Prospering of the Native Americans

Dr. K. Nowall, a scholar who never heard the adage ''Better be ignorant of a matter than half know it'' (Pulilius Syrus), was amazed when he discovered the nearly six-fold increase in the Native American* population from 1860 to 1890. He found that the U.S. Census counts of this minority population were 44,000 in the 1860 decennial census report and 248,000 in the 1890 report [1]. ''This may be the most rapid rate of population growth ever seen, a compound average growth rate of almost 6% per year—phenomenal!'' said the scholar.

Unfortunately, Dr. Nowall did not know the subject matter. If he had known the demography of Native Americans, he would have known that in 1860 most Native Americans lived in Indian Territory or on reservations (lands set aside for these peoples) and they were not included in the U.S. Census counts. ''American Indians were first enumerated as a separate group in the 1860 census. However, Indians in Indian Territory or on reservations were not included in the official count of the United States until 1890'' [1].

B. The Russians Get There First Again

When informed of his inadequate knowledge of the Native Americans, Dr. Nowall turned his attention to the demography of the people customarily called ''the Russians.'' These peoples inhabited the land mass known as the Russian Empire before the revolution of 1917 and afterwards known as the Union of Soviet Socialist Republics (USSR).

He found various census reports which showed an 11% decrease in population in the 13 years from 1913 to 1926, from 166 million people to 147 million. But he failed to recognize that the 1913 figure was an official *estimate* rather than a census count. With his otherwise fertile mind he then found additional census reports which showed a 22% increase in the population in the 20 years from 1939 to 1959, from 171 million people to 209 million.†

These figures were not as dramatic as his discredited figures on the growth in Native American population. But he now had a basis to explain these changes in the population counts in terms of modern demographic, political, sociological, and psychological theories, simultaneously satisfying both russophobes and russophiles. Think tanks on both sides of the political spectrum gave him grants to dig further into these matters.

Dr. Nowall explained the USSR population decrease from 1913 to 1926 in

*This category includes the American Indian, the Eskimo, and the Aleut.

†Raw data from Refs. 2, 3, 4, and 6.

terms of the general downward trend in birthrates which took place in a number of the European countries and the United States during this period. For example, the birthrate in both England and Sweden fell from about 25 per 1000 to 20 per 1000 of population. The birthrate in the United States fell from about 30 to 24 per 1000.* This explanation satisfied those Russians who insist that they are the original inventors, first in all things, and also those Americans who were looking for the USSR to disappear.

As to the increase in population from 1939 to 1959, which took place despite the huge number of Russian military and civilian deaths during World War II, Dr. Nowall explained this in terms of the Russians once again taking leadership, for the national birthrates in most Western countries were on the increase during the postwar period. To avoid losing his russophobic support, he pointed out that the increase in population was undoubtedly deliberately managed by the state to provide a large army for future expansionist ventures.

Unfortunately, Dr. K. Nowall was once again the victim of his own ignorance. Between 1913 and 1926, the Russian Empire and its successor the USSR, experienced World War I, a revolution and accompanying civil war among several factions, foreign invasions and occupations, and a severe famine. In addition, in the peace settlement at the end of World War I, the country lost considerable geographical area (Poland, the Baltic States, and more) along with their populations.

As for the growth in population from 1939 to 1959 (despite the large casualties during World War II), the USSR regained some of its lost territories and their populations at the end of the war. Unfortunately, Dr. Nowall was too young to have lived through these changes and thereby know them from personal experience, and because of slipshod research efforts knew too little about the subject matter.

In truth, when adjustments are made for the changing boundaries, there was no decrease between 1897 (or 1913) and 1939, nor as large an increase between 1939 and 1959 as Dr. K. Nowall imagined [6].

III. Some Examples

A. Music Hath Charms . . .

Dr. Donald D. Atlas, a professor of medicine, found that the mean length of life for 35 male symphony conductors was 73.4 years [7]. Since the mean length of life for the U.S. male population as a whole at that time (1978) was 69.5 years, he concluded that this longevity was the result of the conductors' unusual combination of personal characteristics: talent, driving motivation, and sense of

*Raw data from Refs. 2 and 5.

fulfillment resulting from the recognition they received from colleagues and the public. These characteristics, he argued, lengthened their lives (on the average) beyond that of the other males.

Unfortunately, Dr. Atlas had no knowledge of the general subject, demography, or of studies on longevity. After the article was published, J. Douglas Carroll questioned Dr. Atlas's inference "that involvement with music lengthens life" [8]. The length of life figure of 69.5 years for all U.S. males is the mean length of life, *measured from birth*. But symphony conductors are made, not born. No matter how great their inherent talent, they usually become conductors after age 30. Thus knowing the subject matter of demography and life tables, we must ask this question: If a male survives to (for example) age 32, how much longer is he likely to live? A demographer would say, "What is the life expectancy for a 32-year old male?"*

An American male who survives to age 32 can expect to live an additional 40.5 years, for a total of 72.5 years, according to the 1978 life table [9, Table 5.3]. This value is so close to the 73.4 years computed for symphony conductors that we cannot conclude that there is a real difference. When we consider the quality of the basic data and the many procedures which actuaries apply to them (some of which are discussed in the Appendix) we find it hard to take seriously a difference of less than one year out of 70. Unfortunately for aspiring conductors, there is no basis to claim that music's charms have lengthened the lives of the male conductors in the United States.

B. Using the Space Shuttle to Cross the Mississippi

How a researcher working with computer graphics "discovered" a statistical anomaly gives us another example of how inadequate knowledge of the subject matter can lead to an unfortunate waste of time and talent [10]. The computer researcher and his co-workers used published data on motor vehicle fuel consumption to create computer graphical displays. By careful study of these displays they determined that there had been a change in the definition of motor vehicle horsepower during the period of the data.

But persons familiar with the subject matter would have known that a significant event had occurred: a change in a critical definition. Indeed, the published data which were used by the researchers to create displays from which they deduced the change in horsepower definition had a footnote which clearly stated that the definition had been changed. They should not have had to use computing time, sophisticated software, and display devices to discover a fact that was known to anyone familiar with the field.

---

*We give details of life table concepts in the Appendix.

C. Judge Not, In That Ye Be Judged Yourself

In 1979, a newspaper published a story claiming that women are ''underrepresented'' in the federal judiciary: ''Only 28 of the nation's 605 federal judges are women even though more than 45,000 women are now practicing law in the United States'' [11]. The writer claimed that, since about 10% of the 478,000 lawyers in 1979 were women, at least 10% of the 605 federal judges should be female. Since there were only 28 women federal judges, or about 5%, women were ''obviously underrepresented.''

But a proper evaluation of the claim of underrepresentation is based on the answer to this question: How much time elapses between getting a law degree and becoming a judge? Federal judges are not appointed from the ranks of new law school graduates. From the difference of 13 years in the median ages of lawyers and judges [12, Table 3], we conclude that it is reasonable to assume that, on the average, it is 10 to 15 years before a newly-graduated lawyer enters the ''eligibility pool'' for federal judgeships.

Thus women lawyers eligible for a federal judgeship in 1979 would have had to receive their law degrees prior to 1970. What was the proportion of law degrees awarded to women prior to 1970? Never greater than 4% [13, p. 280]. If you compare this ''eligible pool'' of women lawyers to the 4.6% of federal judges reported to be women, there is no basis to conclude that, in 1979, women were underrepresented in the federal judiciary.

D. Drink Deep, or Taste Not . . . A Little Knowledge Is a Dangerous Thing

F. A. Hassan attempted to estimate the birthrate of ancient populations [14]. He computed what he called the ''total fertility rate,'' defined by him as an estimate of the number of live births which a woman would have if she lived through the reproductive period.

He calculates this total fertility rate by first finding the ''age-specific'' birthrates for the different age groups of women. This is done by dividing the number of births in a given age group by the number of women in that age group. For example, suppose there were 2000 women in the age group 15 to 19 years and that 100 births were reported for the group. Then the ''age-specific'' birthrate for the women in that group is 100 divided by 2000 or .05 (50 per 1000 women), which is the usual way demographers determine such rates.

Hassan found the age-specific birthrates for groups of women throughout the reproductive period, for example 15–19 years, 20–24 years, and so forth. He then added all the age-specific rates and divided by the total number of women in all age groups. This is wrong, of course, since the number of women has already been taken into account in the computation of age-specific rates. While it is impossible to tell exactly how the author carried out his computations, he does say that the age-specific rates are summed ''for mothers of successive age, and the product divided by the number of mothers'' [14, p. 128].

The author's lack of knowledge led him to a misuse, but such knowledge as he had may have been leading him in a correct direction. Perhaps he was trying to compute a meaningful statistic known to population analysts as the ''net reproduction rate.''

The net reproduction rate is a measure of the potential growth of a population which is based on knowledge of births and deaths. Stated briefly (and perhaps too succinctly), it is the ratio of the total number of female births to the total number of women who survive through the reproductive period (from the ages of 15 to 44).*

If the value of the net reproduction rate is 1.0, then the population is just reproducing itself. This means that the survivors of 1000 female births give birth to 1000 females during their reproductive period. If the rate is greater than 1.0, then the population is potentially growing; if the rate is less than 1.0, then the population is potentially decreasing.

For example, out of 1000 females born in the United States in 1979, 955 will survive up to age 45 and 939 up to age 50 years [15]. The net reproduction rate for the United States in 1979 was 94.3 [16, Table 81], indicating a potential decrease of 5.7% in population per generation (about 30 years) in the absence of net in-migration.

E. Primitive Peoples Don't Die Right

Another anthropologist explaining life tables to colleagues implies that the people studied by anthropologists—particularly those called ''primitive''—die differently than the rest of humankind [17]. He gives a number of model life tables in his paper which start with an age for which the life expectancy is 10.7 years. To have a life expectancy of 10.7 years and for the population to survive more than one generation, the birthrate of the population must be 93 births per 1000 of population per year!†

Ninety-three per thousand is an extremely high birthrate. We have rarely seen a rate of over 65 per 1000, which would lead to about nine children per family, even in groups where birth control and abortion are unknown or strictly forbidden. Obviously, when the birthrate is deliberately controlled, the fertility

*The age limits for the group are sometimes stated as 10 to 49 years, but few women under age 15 or over age 45 have births.

†As discussed in the Appendix, life table birth and death rates are equal. Either can be obtained as the reciprocal of life expectancy at birth (one divided by the life expectancy at birth) and then multiplied by 1000 to obtain the conventional rate per thousand.

rate is considerably lower [18, p. 73]. We would expect to find a birthrate higher than 65 only in some special subset of a general population, such as a new housing development occupied only by young couples in their 20s and 30s, or an ethnic or social group in which the women did not work outside of the home.

What are the consequences of a rate high enough to support this anthropologist's life expectancy of 10.7 years? If the population was holding its size and the birthrate was 93 per 1000, then there would be about 200 women in the reproductive ages for every 1000 of population. Since some women are not fertile and some conceptions end in stillbirths, the reproducing women would, on the average, have to have a live birth every other year until they either died or reached the end of the reproductive period. A woman who survived the whole reproductive period would have to bear 14 or 15 children.

Let's look at the facts. In the United States in 1982 there were 53 million women between the ages of 15 and 44, and there were 3.7 million live births. To hold the population at the same size—neither increasing nor decreasing—an average of slightly over two births per women suffices. This is found by dividing 53 million by 3.7 million which equals 14.3, and noting that the reproductive period is 30 years (from ages 15 to 44) and that 30 divided by 14.3 is just over two.

The conclusion is that this anthropologist did not know enough about the subject matter of demography and, in particular, life tables, to serve as a source of knowledge for colleagues; a misuse was committed.

F. The Population Seesaw

Social Security policy is a continuing concern among Americans. Currently, the expected increase in the number of persons eligible for Social Security benefits is stimulating concern among many economists, social scientists, and politicians. They foresee a crisis in Social Security policy by the 21st century because the growing number of recipients will become an "intolerable" burden on the working Americans who pay Social Security taxes.

We continually see and hear these statements, projecting difficult and perhaps insoluble problems, generational conflict, and social upheaval, all due to arise when a "modest number" of working Americans are supporting a "huge older population" drawing Social Security benefits.

One such statement is:

> The percentage of people 65 years old and over increases from 11.4 in 1981 to 13.1 in 2000 and to 21.7 in 2050. According to the . . . authors, an important finding is that the "ratio of the working age population—18 to 64 years—to the retirement age population—65 years and over—declines from 5.4 in 1981, reaches 4.7 in 2000, hits 3.0 to 1 in 2025, and then 2.6 to 1 by 2050. [19]

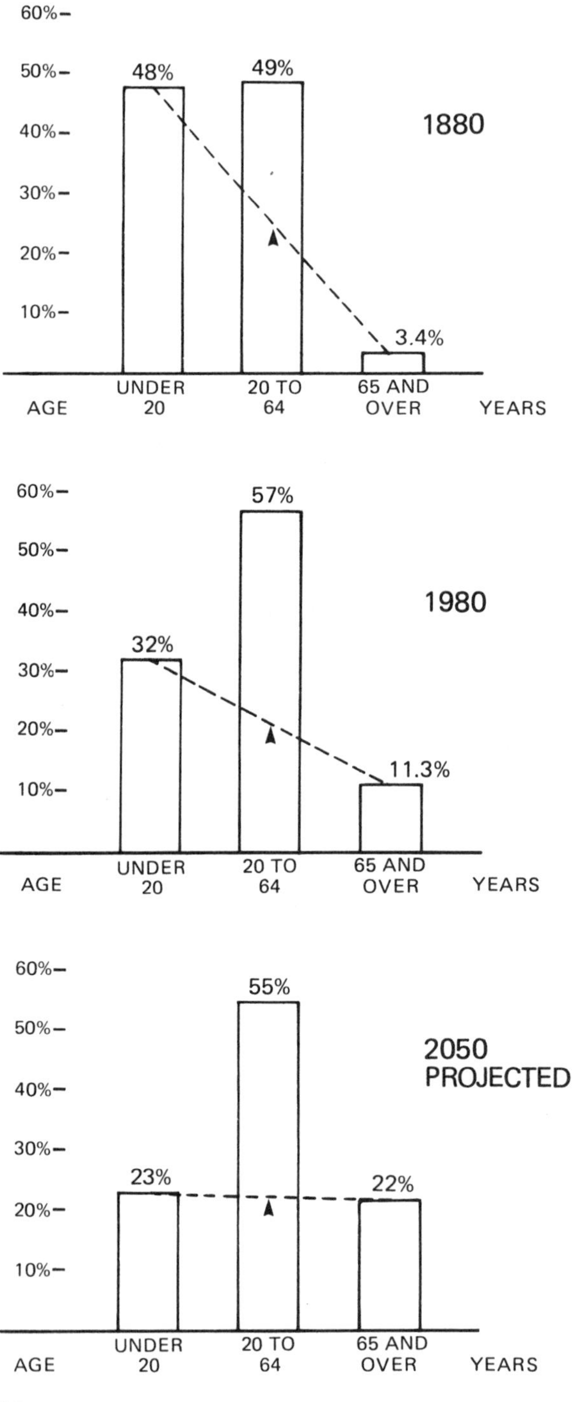
60%
50%
40%
30%
20%
10%
48%
49%
3.4%
1880
AGE
UNDER 20
20 TO 64
65 AND OVER
YEARS
60%
50%
40%
30%
20%
10%
57%
32%
11.3%
1980
AGE
UNDER 20
20 TO 64
65 AND OVER
YEARS
60%
50%
40%
30%
20%
10%
55%
23%
22%
2050 PROJECTED
AGE
UNDER 20
20 TO 64
65 AND OVER
YEARS

Others have made similar statements:

> Low fertility and improved mortality will reduce the ratio of working age persons (aged 20–64) to retirement age persons (65 and over) from 5.0 in 1975 to 3.0 in 2025. Thus, if present trends continue, there will be fewer workers available to pay for the increasing retirement benefits required by the larger older retired population. [20]

U.S. Senator Bill Bradley joined the chorus with a chart which showed that in 1950 one Social Security beneficiary was supported by 16.5 workers but that in 1980 only 3.3 workers were available to support each retired person, and "by 2030 there may be only 2 workers paying into Social Security for each beneficiary" [21].

None of these people (nor others making similar prophecies of doom to come) understand the dynamics of a changing age distribution and how an *entire* population is supported. If these commentators, writers, reporters, and politicians knew the subject matter, they would know that age distributions tend to resemble seesaws with fulcrums being the population aged from 20 to 64, as we show in Figure 3.1. If the proportion that is younger decreases, then the proportion that is older increases, and vice versa. The workers, who are mostly aged 20 to 64 years, support the younger population as well as the older population. Children are not self-supporting; both they and the older Social Security beneficiaries are supported by the working population. When commentators compare only those age 65 and over with the working population, they commit a misuse of statistics, either through insufficient understanding of the subject or through the desire to make a political point.

Throughout the last century and perhaps further back, about half of the total population was aged 20 to 64—the prime working force ages—and this half of the population supported the other half of the population, the dependents.

What is the trend today in the age distribution of dependents? The birthrate of the United States has decreased significantly over the past several decades; the proportion of the population that is younger has decreased and the proportion that is older has increased. The 1980 dependent proportion was 43%, the Census Bureau's projection for 2050 is 45% [22]. The change in the dependent proportion is less than the forecast error! The seesaw works: The projection is that the proportion of young dependents will decrease as the proportion of older dependents increases.

Thus, the issue is not what some commentators and politicians claim: the numbers of persons age 65 and over. Since the national will is to support

---

Figure 3.1 The seesaw of dependency. The dependent proportion of the population stays approximately constant, but the proportion of dependents is shifting from lower to higher ages as shown by the balance of the dotted seesaw line. (From Refs. 22, 23.)

dependents, the issue is: By what mechanism will the funds for their support be distributed? The population in the working age class will be contributing about the same amount to support the dependents under age 20 and age 65 and older. It has been the custom for most (not all!) of the support of dependents under 20 to be paid by the working age class as parents. But most of the contributions to the support of the dependents age 65 and older have been paid as Social Security taxes contributed by both current workers and the ultimate recipients when they were in the work force. As the relative proportion shifts (as shown in Figure 3.1), the mechanism of the funds transfer may change.

We are not the first observers to recognize the seesaw of dependency. In 1978, Robert L. Clark and Joseph J. Spengler made this observation in an analysis that is an excellent example of a proper use of statistics [23, Chapter 2, Changing demography and dependency costs, pp. 50–90]. As we do here, they show the seesaw effect of the changing age distribution of the United States. Since they made their observation, the U.S. Census Bureau's population projections have changed, but the seesaw remains.

We are left with some troubling questions. Why did politicians and some experts misinterpret the population numbers about the changing age distribution of the United States in a way that made the increase of older dependents appear to be an "intolerable" burden on working Americans who pay for Social Security? It might have been no more than lack of knowledge of the subject matter.

IV. Summary

Numbers do not interpret themselves. A number is a number—no more and no less. It can make sense and be understood only in light of some particular subject matter. The writer or analyst must know the subject matter before the numbers can be explained.

Beware of anyone who simply shoots out numbers—unless you are bidding at an auction, when you should be extra careful.

4
Definitions

If you wish to converse with me, define your terms.

Voltaire

He that can define . . . is the best man.

Emerson

I. Introduction

Crime soars. Poverty declines. High school dropout rates reach an all-time peak. Unemployment goes down. Trade balances go up. The number of speeders increases. These are among the many simple statements about crime, poverty, unemployment, and other social phenomena which affect our daily lives. All of these statements are without substantive meaning unless supported by statistics, as they usually are.

But what is that meaning? To know for sure, you must find out how a general term (i.e., "crime," "poverty," "unemployment") is defined. For no matter how much effort we invest in collecting, refining, or improving the data, the resulting general conclusion is based on a statistical definition. Any statistic we compute is an artifact.

For example, what does "crime" mean? What is the definition of the measurement used to measure crime which is "soaring"? Is it the increase in the offenses reported to the police? Or is it the increase in victimization rates? How much credence shall we give to either measurement? To answer this, we must dig even deeper: What is an "offense reported" and what is a "reported victimization"? How do we define these terms?

> One perspective [in determining the extent of crime] is provided by the FBI through its *Uniform Crime Reporting Program* (UCR). The FBI receives monthly and annual reports from law enforcement agencies throughout the country, currently representing 97 percent of the national population. Each month, city police, sheriffs, and State police file reports on the number of index offenses [murder and negligent manslaughter, forcible rape, robbery, aggravated assault, burglary, larceny, motor vehicle theft] that become known to them.
>
> The monthly Uniform Crime Reports also contain data on crimes cleared by arrest. . . .
>
> *National Crime Survey* (NCS)—A second perspective on crime is provided by the NCS of the Bureau of Justice Statistics. The NCS includes offenses reported to the police, as well as those not reported to police. . . . Details about the crimes come directly from the victims. No attempt is made to validate the information against police records or any other source. The NCS measures rape, robbery, assault, burglary, personal and household larceny, and motor vehicle theft. Murder and kidnapping are not covered. Commercial burglary and robbery were dropped from the program during 1977. The so-called victimless crimes, such as drunkenness, drug abuse, and prostitution are also excluded, as are crimes for which it is difficult to identify knowledgeable respondents or to locate data records. [1, pp. 163–4]

The several statistics—the Uniform Crime Survey, and the National Crime Survey—may be in conflict, as you can easily see if you know the definitions. Which statistic do you use? It's impossible to answer this question out of context. When you know the purpose of your analysis, you can choose the most useful statistic for that purpose.

In this chapter we give examples of misuses of statistics involving definitions.

II. Preliminaries

A. Who's Who?

Do you define a "manager" by job title? If so, you include persons with responsibility for a part of a small office along with persons with responsibility for billions of dollars in projects. Many job titles given for occupations do not necessarily match salaries, responsibilities, and authority. But more importantly, in many job situations, a person's title is based only upon his/her self-description. Most of our information on occupations comes from surveys such as the Census and the Current Population Survey, in which the respondent is asked to name his or her occupation. Sometimes there is no problem: A bookkeeper is a bookkeeper. But what about the head of the accounting department which includes the bookkeeper? He or she can report "accountant" or "manager." These self-reported designations are then classified into whatever formal system is in use at the time. In addition, the classification system often changes to fit the needs of the moment. Since the apparent needs are always changing (we no longer list carriage or harness makers as occupational categories), the occupational classification system is always being reexamined and revised as deemed necessary. As noted by the U.S. Bureau of the Census:

> Comparability with earlier census data—Comparability of industry and occupation is affected by a number of factors, a major one being the systems used to classify the questionnaire responses. . . . The basic structures were generally the same from 1940 to 1970, but changes in the individual categories limited comparability from one census to another. These changes resulted from the need to recognize the "birth" of new occupations, and the "death" of others. . . . In the 1980 Census . . . the occupation classification . . . was substantially revised. [2, Appendix B, p. 8]

"Definition" and "classification" are the twin names of the game. Watch both.

B. You Are What You Say You Are

Race is another concept for which there are no natural or correct definitions. We are not even sure what race means. If race means homogeneity of genetic structure, then we are indeed in deep trouble. It is likely that every person alive is a hybrid in the genetic sense. And we do not have the genetic code available when we ask individuals to report their own perception of their race, which is the usual way to obtain data on race.

In the United States and some other countries, skin color is the basis for classification of race. Thus, many people listed themselves as "brown" in the 1980 census. This category does not appear in the official census list. The designation "white" is also suspect: As Mark Twain pointed out some time ago, most so-called white people are really pink!

Skin color is a social definition of race, as you can see from the history of race definition in America. The U.S. Constitution specified that congressional representatives were to be allocated according to a decennial census, in which the categories were defined as follows: "Representatives . . . shall be apportioned among the several States . . . according to their respective numbers, which shall be determined by adding to the whole number of free persons, including those bound to service for a term of years, and excluding Indians not taxed, three fifths of all other persons" [3].

Since the slaves were "black," the Constitution defined three "race" categories for census purposes: free persons (which could include a small number of blacks and all indentured servants, and those few taxed Indians); untaxed Indians; and "other persons" (who were mostly black slaves). As the nation grew and changed its social structure, those social changes affected the census categories and concepts of race. In the XIVth Amendment to the Constitution, the categories for apportionment were reduced to two: ". . . counting the whole number of persons in each State, excluding Indians not taxed."

By the middle of the 19th century the decennial census had become a source of general information. Hence, the racial classification of white and black was retained. By 1860 enough Chinese persons had immigrated to the United States to warrant adding the racial category of Chinese to that census. In 1870, following the immigration of Japanese persons, that category was added to the census racial classification system. Anyone who claimed not to be white, black, or American Indian was given a separate "race" designation. The 1980 census schedule lists the following categories: "White, Black or Negro, Japanese, Chinese, Filipino, Korean, Vietnamese, Indian (American), Indian (Asian), Hawaiian, Guamanian, Samoan, Eskimo, Aleut, and other." The respondents make their own choices from this list. Thus persons who view themselves as "Brown" could check "other." Others are counted as "race, not elsewhere classified," and the 1980 census lists 5.8 million Americans who chose this classification [2, Table 74].

The number of Hispanics in the United States increased greatly after World War II, and since there was no simple way to include them into a racial classification, a separate question was added to the schedule to determine whether the respondents see themselves as Hispanic. Published tables carry this category separately from those of the various races.

C. Has There Been a Change or Hasn't There?

1. *In Prices?*

Definitions in transition often create headlines that mislead. "Producers' Prices Show Slight Drop, First in 4½ Years," reads the headline [4]. "The decline was centered on motor vehicles, food and energy," says the text. But the final paragraphs of the article reveal that there was a change in definition which caused the drop. If the definition had not been changed, the Index of Producers' Prices would have shown a rise of twice the magnitude of the decline noted in the headline.

What was it about the change in definition that caused the "slight drop"? It was using a definition of automobile price that now included the discounts given on new automobile sales at the end of the model year. In the year before the change in definition, the price of automobiles was entered into the index according to their "sticker," or list, price. If you haven't bought a car recently, the "sticker" price is the price stated on the sticker that comes on the car from the factory. It is customary in car purchases for sales persons to negotiate a lower selling price, based on several factors including the value of the trade-in car, the accessories purchased, and a discount. The discount tends to rise at the end of the model year when dealers are anxious to reduce their inventories in preparation for the arrival of the new models. However, most automobiles are not sold at this ''sticker'' price—they are sold at a lower, or ''discounted,'' price.

To illustrate the effect of this change in definition, suppose that the average automobile sticker price for a given year is $10,000. If, in the computation of the index of producers' prices, no account is taken of discounts, the figure of $10,000 is added to the prices of other items which comprise the index. However, if an average discount of 20% is offered in the last quarter of the year, then the average automobile price added to other prices in the index would be $9500. This average price is based on an average price of $10,000 for the first three quarters and 20% less, or $8000 for the last quarter—three-fourths of $10,000 plus one-fourth of $8000. Thus, the automobile price amount entering the index would be $500 less than under the original definition.

The discounts become greater as the model year draws to a close since dealers want to sell off those cars from the current year which will soon be replaced by new models. The revised index includes the prices of cars at an estimated discount, rather than at sticker price as in previous years, thereby lowering the Index of Producers' Prices.

This drop did not represent a change in producers' prices; it represented the change in the definition of ''price.'' Thus we can see that ''correctness'' is arbitrary in the statistical context.

2. *In the Number of Poor People?*

We can correctly (or incorrectly) count the number of families below some arbitrary poverty line, but we cannot correctly determine the number of poor, for:

> Families and unrelated individuals are classified as being above or below the poverty level using the poverty index originated at the Social Security Administration in 1964 and revised by Federal Interagency Committees in 1969 and 1980. The poverty index is based solely on money income and does not reflect the fact that many low-income persons receive noncash benefits such as food stamps, medicaid, and public housing. The poverty index is based on the Department of Agriculture's 1961 economy food plan and reflects the different consumption requirements of families based on their size and compositions. The poverty thresholds are updated every year to reflect changes in the Consumer Price Index. [1, p. 429]

3. *In the Hourly Earnings in the Aircraft Industry?*

The U.S. Bureau of Labor Statistics' (BLS) monthly earnings survey contains a line for the average hourly and weekly earnings in the aircraft industry, but that line is empty [as shown in 5, p. 1888]. There are no entries

> because the bureau [of Labor Statistics] and the aerospace industry cannot agree on how the number should be calculated.
>
> The dispute presents a picture of a major industry saying it was blindsided by

> a bureaucracy's refusal to adjust its record keeping to reflect what's happening in the real world. Industry representatives are even threatening to hire their own private data gatherers to produce the missing number, which the BLS has not published since October 1983.
>
> The BLS, a Labor Department agency, is sticking to its position, saying it won't change its definition of what's in its numbers simply to satisfy an industry's contractual troubles. "We just won't change definitions in midstream," said BLS commissioner Janet L. Norwood. [5, pp. 1888–1889]

This dispute over a definition started in 1983, when major aerospace companies gave once-a-year lump-sum payments to workers rather than wage increases, a practice which continues. Lump-sum payments are income to workers, but they are not worth as much as wage increases because the amount is not added to the workers' pay in figuring overtime pay, vacation payments, and fringe benefits. The Bureau of Labor Statistics took the position that the lump-sum payments should not be added into the reported hourly and weekly earnings numbers because it is a once-a-year event and thus does not fit the bureau's definition of hourly and weekly earnings.

When asked for the usual monthly reports on workers' earnings, the aerospace companies included the pro-rated lump sum payment. The Bureau of Labor Statistics refused to use these data. Hence, the blank line in the average hourly and weekly earnings report for the aircraft industry.

What is the consequence of this dispute over a definition? There are three parties to whom the missing numbers are important: the aerospace companies, their customers, and their workers. The interests of these three parties conflict—the companies want the lump sum payments included so as to get reimbursed for their payments and for protection against rising costs; the customers of these companies would prefer not to include these payments so as to reduce the cost of purchased aircraft; and the representatives of the companies' organized workers either want the lump sum payments made part of the workers' earnings base (hence increasing overtime payments) or not counted at all. There is also a question as to whether the lump sum payments should be retroactively applied to contracts already in force; the aerospace companies' customers (through the Air Transport Association) have agreed to include them in future contracts, but not retroactively.

The U.S. Department of Defense is a major customer of the aerospace companies and has received its share of blame for the conflict from the representatives of organized labor ". . . for putting pressure on the aerospace companies to cut their labor costs with hardline negotiating tactics, leading to the companies' current predicament" [5, p. 1890].

This dispute over a definition is not just a semantic squabble. If the price uncertainty is about one percent, then the amount of money in dispute on a five-billion dollar contract is about fifty million dollars. The Defense Department writes many contracts for more than five billion dollars; the tens of millions of dollars involved in potential price uncertainty could add up to real money.

III. Numbers and Conclusions in Conflict

Is the stock market rising, falling, or static? When we ask this question we are trying to determine the general movement of stock prices: whether the market is a bull, bear, or sloth market. To answer this question, we form indexes by combining selected stock prices. These indexes are usually in agreement about market movement. This congruency reinforces our faith in the value of the indexes. But what would we think of these indexes if the Dow-Jones, Standard & Poor's and American Exchange indexes were in conflict about the market movements day after day?

If these statistics were in continual conflict, then we would cease to accept their conclusions and would demand their revision. Unless we were convinced that one was superior to all the others, we would draw no conclusions about market movement.

Let us look at some of the cases in which we have conflict among statistics which are supposed to measure the same entities; conflict so serious that we must question our ability to use these statistics in the formulation of policy.

A. A Case of International Discord

Comparisons of the United States with other countries are often made with respect to such vital statistics as the birthrate, infant mortality rate, income, unemployment, etc. But all countries do not use the same definitions or methods in collecting data. Hence, such comparisons are dangerous and full of traps for the unwary. We cannot review the procedures of all countries, but can only—as we do elsewhere—give an example to show you just how dangerous it is to accept these comparisons without question. When you hear of, or make, such comparisons, we advise you to study carefully the technical appendixes and introductions to the books containing the published information.*

The introduction to the *United Nations Statistical Yearbook* has the following warning:

> One of the major aims of the *Statistical Yearbook* is to present country series which are as nearly comparable as the available statistics permit. . . . There are certain limitations . . . of which the reader should be aware. For example, a direct

---

*Herbert Jacob has written an excellent, if somewhat technical, guide for such evaluations. See Herbert Jacob, Using published data: Errors and remedies, *Sage University Paper Series on Quantitative Applications in the Social Sciences, 07-001.* Sage Publications, Beverly Hills and London, 1984.

> comparison between the official national product data of countries with market and centrally planned economies would be misleading owing to differences in the underlying concepts. [6, p. xviii]

When the above statement is translated into straight talk, it says that countries with market economies (such as the United States) and those with planned economies (such as the USSR) define economic quantities differently. While you might be able to make adjustments to improve comparability, you still must make comparisons cautiously, *very* cautiously, to avoid misusing statistics.

Another example of noncomparable definitions comes from another United Nations publication:

> Most of the vital statistics data (births, deaths, marriages, and divorces) published in . . . [the *United Nations Demographic*] . . . *Yearbook* come from national civil registration systems. The completeness and accuracy of the data which these systems produce vary from one country to another. . . . . [7, p. 13] A basic problem facing international comparability of vital statistics is deviation from standard definitions of vital events. An example of this can be seen in the cases of live birth and foetal deaths. In some countries or areas, an infant must survive for at least 24 hours before it can be inscribed in the live-birth register. Infants who die before the expiration of the 24-hour period are classified as late foetal deaths and, barring special tabulation procedures, they would be counted either as live births or as deaths. But, in several other countries or areas, those infants who are born alive but who die before registration of their birth (no matter how long after birth registration may occur) are also considered as late foetal deaths. [7, p. 15]

The following simple definition of a live birth is internationally agreed to and defines the birth which is used in computation of the birthrate: If the newborn shows any evidence of life it is counted as a live birth even if it lives for only a few moments. Whenever a person dies—at the age of one day or less, or at the age of 100—it is registered as a death. Since some countries deviate from this definition of a live birth for statistical purposes, you should make international comparisons of birth-related data only when you know the precise definition used by the countries you are comparing.

It is not only for demographic data collection that different nations use different definitions for ostensibly the same statistic. Examples in manufacturing, immigration, medical administration, unemployment, and so forth, can be given. One example concerning accounting for exports and imports appears in Section E, below. But the careful reader is rarely at a disadvantage, for national and international statistical compilations usually give definitions and cautions at length. In cases where definitions are not given explicitly, it is usually possible to track them down. Problems arise when the headline writer, the Dr. Nowall, or the politician, in an effort to make a point, will omit, accidentally or deliberately, such cautions from their statements.

B. A Case of Statistical Truancy?

The choice of a definition can have a great effect on the resulting statistics. Where the statistic is used to help set public policy, a misinterpretation can become a key factor in setting a policy. The consequences are rarely good.

Albert Shanker, head of the United Teachers Federation, caught New York's new chancellor of public schools in a misuse which illustrates how definitions can be used to further political ends. The 1979 "Mid-Year Report" on the public schools released by the chancellor's office reported that the dropout rate of high school students was shockingly high. The report stated that there was a 45% dropout rate, a figure which merited national media coverage. According to Shanker, the Report did not "establish a 45% dropout rate" for New York City schools. It arrived at an *estimate* and stated: "The estimate cannot be precise because of the manner in which the system assembles the pertinent data; the careers of individual students are not followed and enumerated" [8]. The report says that 85,459 students entered high school in 1974 and that 48,173 graduated in 1979; the difference, 37,286, is considered to be, and is defined as, "dropouts."

But is this definition correct for the purposes at hand? The report assumes that *all* students, those graduating both public and private eighth and ninth grades, entered the *public* high schools. About 15% of the students leaving eighth and ninth grades are in private or parochial schools and many of these students may enter high schools outside of New York or leave the city. About 13,000 of the missing 37,286 were enrolled in an Evening Certificate Program. An "Auxiliary Service Program" which serves students in a work-study program had 15,526 enrollees and graduates, who were counted as dropouts in the Report. Is this a deliberate or accidental misuse? At the time the then-new chancellor was negotiating with the city administration for funds. Shanker believed that the governmental purpose was twofold: first, to gain support for increased funds; and second, to enable the chancellor to show an improvement as a consequence of his subsequent actions [8].

C. When Is a Dispute Not a Dispute?

A dispute may not be a dispute when the conclusions are based on different definitions. The headline reads "Study Disputes Earlier Questions on Visits to Prisons by Youths to Reduce Jersey Crime," but what is the real story? The article states: ". . . a Rutgers University study . . . showed that juvenile visitors who were subjected to verbal assault by hardened convicts in a . . . visit to the prison actually committed more crimes afterward than did a control group" [9]. However, a subsequent study by researchers at Kean State College found that: ". . . the behavior of 47 percent of an experimental group of 66 juveniles (compared to the behavior improvement of only 26% of the control group) . . . had improved after visiting with the convicts'' [9].

Does the program (called ''Scared Straight'') have a good or bad effect? It appears that one study proves yes, the other no. The Kean State College researcher says that it is difficult to compare the two studies because of differences in the way that the two groups of youth were defined: (1) about 50% of the juveniles in the Rutgers study had police records compared to 100% for his study; and (2) the Rutgers study covered a six-month period after the visits whereas his study covered 22 months.

But even if these factors are removed, how can we compare ''improvement in behavior'' with ''committed crimes''? As reported, we have no way of knowing how these two critical concepts are defined. In the absence of further information and clear definition of terms, we do not know how to evaluate these studies for accuracy.

D. Is Anyone Fooled?

The Japanese balance of trade is important in international policy making, and the higher the surplus shown by Japan, the greater the pressure on that country to modify its policies. Presumably, we can get the trade surplus or deficit by subtracting the value of imports from the value of exports.

But what is ''value''? The purchase cost is certainly the major element in the value of either imports or exports, but what about shipping, stevedorage to the shipping vehicle, insurance, and local shipping to the export vehicle? These several components are the basic inputs to an explicit definition of import or export value.

Japan, which has the most to gain by reducing the apparent value of its trade surplus, adds insurance and all freight charges to its imports, raising the value of imports. However, only local transport and stevedorage are added to the value of exports. The U.S. Commerce Department defines value identically for both imports and exports as the total cost at the vehicle of shipping, less stevedorage. The International Monetary Fund subtracts about 10% from the value of Japanese imports and recalculates the net trade balance.

We all know that there is a high relative error in a value obtained by subtracting two large numbers which are close together. Thus, it is not surprising that the reductions in trade surplus with Japan reported by the United States and Japan often differ by a factor of two. For example, the Japanese trade surplus with the United States in November, 1978, was about .8 billion dollars as estimated by the Japanese and about 1.6 billion dollars as estimated by the U.S. Department of Commerce [10].

It is much easier for someone who is well informed in the practice of international shipments and accounting to identify misuses and avoid being misled. In this case, the expert can use his knowledge to ''adjust'' the reported statistics to obtain comparability. Unfortunately, it is not always possible to make statistics comparable by adjustment, especially if the reader is unfamiliar with the subtleties.

But a lay person can, by asking questions about definitions, get sufficient information to avoid a misuse. To evaluate the three measures (Japan's, the International Monetary Fund's, and the U.S. Department of Commerce's) of Japan's trade balance, you do not need to know in advance that Japan evaluates its imports on a C.I.F. basis (cost, insurance, freight) and its exports on an F.O.B. basis (free on board, cost, and all costs to get to the ship, such as stevedorage), or that the United States Commerce Department evaluates both on a free-alongside-the-ship basis (F.O.B. less stevedorage). Once you see the discrepancy, you know that you must check further and find out how the three agencies calculated the trade balance, and if the definitions vary you either seek informed help or forget the numbers.

E. TB or Not TB?

A time series is a set of data taken at different times during some period. Typical time series in government are: the shipments of automobiles, net trade balances, population, corporate profits, time deposits, and insect infestation. All of these may be reported by day, week, month, quarter, year, or decade. We are concerned with the problem of comparability that arises because of changes in definition. Take, for example, the following case, which was reported in the *Washington Post*: "The tuberculosis rate in the District (of Columbia) . . . dropped 30 percent last year to the lowest rate in 30 years, Mayor Marion Barry announced yesterday. . . . The number of reported cases in the city dropped from 341 in 1980 to 239 in 1981" [11]. It is quite possible, as the director of the district's TB Control Program said, that the decline was due to the agreement in 1981 of physicians and clinics to use a new double-drug therapy. However, our evaluation of whether there was such an effect is clouded by the change in definition of a "reported" TB case. In 1980 "all suspected tuberculosis cases reported to the TB control clinic by doctors and health clinics were included in statistical reports unless tests specifically rule out the disease. . . . But in 1981, only medically confirmed cases were included."

It is not a misuse of statistics to change the definition of a statistic from one period to the next. In fact, there are cases where it would be a misuse not to do so. However it *is* a misuse to fail to recognize or state that a drop or increase in some important value could be due to a change in definition.

IV. Can It Be Done Right?

The answer is yes, and it can be done in a newspaper. In a story entitled "How many emigrants are there really?," *Jerusalem Post* reporter David Krivine made a laudable attempt to surmount the difficulties of definition and answer the question [12]. In so doing, he avoided the temptation to write the attention-getting headline "One in five Israelis flees the country," and gives us a model of careful statistical analysis reported by a nonstatistician.

What data were immediately available to Krivine? One newspaper reported a figure of 700,000 emigrants without a source. The director general of the Jewish Agency estimated the number of Israelis settled in the United States as between 300,000 and 500,000, hence the number of Israeli emigrants in the world must be something larger.

Krivine did not take the easy way out and pile estimate on estimate. He turned to the Israeli Central Bureau of Statistics (CBS). Since 1948 the Bureau has counted the number leaving Israel ("out-migrants") and the number returning ("in-migrants"). The net out-migration computed from these data (without examining definitions) from 1948 to 1984 (out-migrants minus in-migrants) is about 340,000. Assuming that there is no illegal out- or in-migration, this amount is the upper limit to emigration, for as shown below, the number of out-migrants is undoubtedly smaller than this number.

This situation begs for definitions. How do you define an out-migrant? Is it someone who leaves the country and returns after one year? Two years? Three years? Or never? And how do you count temporary residents? For example, if a Canadian moves to Israel, lives there for a few years and then returns to Canada, is this person an in-migrant, an out-migrant, or neither?

The CBS collects data about individuals who leave Israel and stay away for more than four years, which Krivine feels "is a fair description of an emigrant." The CBS started counting these individuals in 1965, and from that time to 1984, the number of persons who left the country and stayed away for four years is estimated to be 151,000.*

Krivine does not attempt to get a "true" count, which would call for a much more extensive analysis. But what he does is use good statistical reasoning about definitions to show that the true count must be much smaller than most of the numbers being bandied about in attempts to influence public opinion both inside and outside Israel.

V. Summary

For the reader:

> Take nothing for granted. Find and understand the definition of key statistics given in the article or report.

---

*The CBS used the actual count for persons who left before 1980, and estimated the number of returnees among those who left after 1980 "on the basis of trends."

If no definition is given, don't give credence to the results reported until you get one. If the issue is important to you, contact the author and get a definition.

If you can't understand the definition, seek an explanation, either from the author or from other sources.

If the definition you get doesn't make sense to you, don't accept the results.

If two or more statistics are compared, check the definitions for comparability. Be persistent in seeking out differences and the methods of adjustment.

For authors:

Give definitions in your reports and articles. And when comparing two statistics, give full information about differences in definitions and the nature of adjustments.

5
The Quality of Basic Data

Her taste exact
For faultless fact
Amounts to a disease

W.S. Gilbert

Measurement does not necessarily mean progress. Failing the possibility of measuring that which you desire, the lust for measurement may, for example, merely result in your measuring something else—and perhaps forgetting the difference—or in ignoring some things because they cannot be measured.

George Yule

I. Introduction

"Garbage in, garbage out" is the rule in data processing. It is also true in the use and misuse of statistics, as we show in Table 2.1. Even if the "garbage" which comes out leads to a correct conclusion, this conclusion is still tainted as it cannot be supported by logical reasoning and is, therefore, a misuse of statistics. Obtaining a correct conclusion from faulty data is an exceptional case. Bad basic data (the "garbage in") almost always leads to incorrect conclusions, and often, incorrect or harmful actions.

It is hard to get good basic data, even in simple, noncontroversial situations and with the best of intentions. Consider the following example that often arises in practice.

U. Lissy, Dr. Nowall's uncle, runs a fast food store. A business consultant, whom Lissy consulted when the store began experiencing problems, suggested that he collect data on the flow of patrons as basic information for a plan for future operations. ''No need to collect data,'' says Lissy, ''I've been running this shop for several years and I'm getting a good 20,000 customers a week. Divided by the number of days a week the store is open (seven) this comes to 2857 per day, and that's a pretty darn good number.''

But is this number ''pretty darn good''? The appearance of the figure of 2857 per day implies counting for a precise value (not 2856, not 2858, but exactly 2857), and it might well be accepted as a precise value and used as such in future calculations and analysis. However, this value has its origin in a guess: it is a ''mythical number,'' and, as Max Singer has pointed out [1], mythical numbers can have an amazing vitality, persisting in the face of all future evidence (see, for example, Section II of this chapter and Chapter 13).

The consultant is both dubious and cautious, however. He persuades Lissy to count the patrons. As the owner, Lissy doesn't want to stand at the door and do nothing but count patrons for a week, so he hires a high school student to count the patrons. The student counts conscientiously the first day, but soon becomes bored and is frequently distracted by conversation and the passing scene. When distracted, the student either records none of the entering patrons or guesses a number for the periods when he/she is not actually counting. The result is an inaccurate number, which is simply wrong. But how wrong? Too high or too low, and by how much? Five percent may not matter for some purposes, but for ordering or planning future purchases, a 5% error may have serious economic consequences for Mr. Lissy. Whether the data are bad enough to give ''garbage out'' depends on our purpose for making the count.

Once you ask ''What is the purpose?,'' the question of what is good basic data becomes more complex. What does Mr. Lissy want? Does he want to serve his present clientele better? In that case, does he have enough support services (clerks, checkout stations, cash registers, and so on)? Is he ordering and preparing enough to meet demand? Or does he want to increase patronage? In that case, can he serve more people efficiently?

We started with the vague concept of ''counting patrons''—people who purchase food at the store. But should that count include children accompanying adults? A group of customers, only one of whom makes the purchase for the group? The person who wanders in to get change for a phone call?

If the goal is to analyze the process of service (clerks, checkout), then the owner may be concerned only with the number of purchasers, and to get that value he can count the number of sales tickets. But if the owner is concerned about marketing, then he needs to know if his services will be sufficient for increased numbers of customers and how much he has to increase his food purchases and preparation services to meet the extra demand. In this case, he may want to analyze the purchases of single people versus families and try to ascertain, by count, the number of singles and families. If his goal is to evaluate the adequacy of facilities (rest rooms, floor space, door openings, and so forth), total count rather than sales tickets may be more valuable.

Thus, even if we had a diligent, accurate counter, the data may be good for some purposes and bad for others: ". . . virtually all basic statistical data are artifacts" [2], and are only correct within a particular context.

In this chapter, we use examples to explore the many types of bad data and their influence on results.

II. Mythical Numbers

A. Incorrect Numbers of Unknown Origin

The Federation for American Immigration Reform (FAIR) distributed a letter of solicitation for contributions which said:

> Immigration helped build this country, but immigration that is out of control could destroy it. . . . Continued immigration at current levels—800,000 legal immigrants in 1980 and perhaps an even higher level of illegal immigrants—will add more than 70,000,000 people to the U.S. population in just 50 years, with no end to growth in sight. [as quoted in 3]

The U.S. Immigration and Naturalization Service (INS) defines (legal) immigrants as nonresident aliens admitted to the United States for permanent residence. The INS includes in that category persons who entered the United States as nonimmigrants but who subsequently change their status. The number of legal immigrants in 1980 was 531,000 [4, p. 83 and Table 124], not 800,000 as incorrectly stated by FAIR (which has a financial stake in telling potential contributors that the number is large).

Furthermore, many immigrants return to their home countries and others die while in the United States. Thus, the number alive and living in the United States is necessarily smaller than the official number of admitted immigrants.

Are the official data better than those of FAIR? FAIR gives no source for its value of 800,000. The official figure is based on permits granted. Such counts are subject to errors in counting, tabulation, transmission of data, and incompleteness. But even if you take a skeptical view of the data collection process, it is hard to imagine a total error greater than 10%. And you don't know in which direction the errors might be! At best, FAIR's figure of 800,000 is way off.

B. A Taxing Example: A Stew of Incorrect and Correct Numbers

In a speech on February 5, 1981, President Reagan made statistical statements which were important to the formation of public tax policies. Although no source for the data is given, some of his comments can be verified within rounding error by data appearing in the *Statistical Abstract of the United States*.

In particular, the President said: ''Prior to World War II, taxes were such that on the average we only had to work just a little over one month each year to pay our total Federal, state and local tax bill. Today we have to work four months to pay that bill'' [as quoted in 5]. Is this true in general, or has the President exaggerated the rise in taxes? To get comparable figures, we convert the President's statement to percentage growth in working time needed to pay ''that bill.'' The three-month increase from one month to four months is a 300% increase.* Data on the amount of direct taxes (individual income taxes, sales and gross receipts taxes, and property taxes) are given in official government publications [7–10] in aggregate form. From these data we can calculate the values of taxes paid ''on the average.'' Table 5.1 shows the parts of direct total taxes, averaging the values for 1934, 1936, and 1938 to get a value for ''Prior to World War II'' and using the data for 1977 for ''Today.''

The average total personal income in the United States for the years 1934/1936/1938 was $59.3 billion, and for the year 1977, $1518.4 billion [8, Table 352 and 9, Table 724]. Thus, total direct taxes have risen from 14% of personal income to 21%. The mean number of months required to pay these taxes has risen from 1.7 to 2.6 months. This 53% rise is significant, and unquestionably painful, but it is about one-sixth of President Reagan's figure of 300%. We could find no data to support the President's value.

In the same speech the President stated that the percentage of earnings taken by the Federal government has doubled since 1960; the official data support an increase of no more than 12%. Since statistical data given in the same

Table 5.1 Direct Total Taxes, in Millions of Dollars

Year	Individual income taxes	Sales and gross receipts taxes	Property taxes	Total direct taxes
1934/1936/1938	933	3,363	4,203	8,499
1977	185,970	83,775	62,535	332,280

Sources: For the years 1934/1936/1938, Reference 7, Series Y384–400; for 1977, Reference 10, Table 477.

*Anyone can create a misuse, and we have done it ourselves. The authors are indebted to Mark L. Trencher for correcting our two errors in our original discussion of this topic [6]. As you might expect, our errors were in the direction which increased the impact of the example as a misuse. Fighting misuses of statistics is a lifetime occupation.

speech concerning population growth agree with official sources to within fractions of a percent, we must conclude that the data on taxes are false and of unknown origin.

C. Data Without Sources

1. *How Many Illegal Aliens?*

When data are reported without a source, you can't take them seriously until you find a confirming source. Again and again, we find attention-getting statements which test credibility, given without sources. There is no shortage of examples of such misuses of statistics in headlines. For example, the headline of a staff-written article appearing in the *Denver Post* in 1980 says "20 Million Illegal Aliens Get Reprieve," and the article goes on to say that "Most of the estimated 20 million illegal aliens in the United States have been granted a reprieve from arrest, prosecution and deportation through a national policy decision made by the U.S. Census Bureau" [11].

Counting illegal aliens is not easy and the purpose of the Census Bureau's policy was to improve the estimate of the number of illegal aliens by reducing their resistance to identifying themselves for counting purposes. David Crosland, acting Immigration Naturalization Service commissioner, said the basis for the policy was "to insure that census operations take place in an atmosphere conducive to complete participation and disclosure of information by all groups" [11].

What of the claim of the article in the *Denver Post* that there are 20 million illegal aliens living in the United States in 1980? Is it possible that about 9% of the U.S. population is comprised of illegal aliens? To some this may sound like a remarkably high proportion of the U.S. population, but perhaps even this number is low. Or could it be a reasonable estimate? Unfortunately, the article provides no source for this number. Thus we don't know who supplied this number or how it was obtained. How can we give it any credence?

Jeffrey S. Passel, chief of the Population Analysis Staff of the Population Division of the U.S. Census Bureau, gives us a credible estimate in a paper he presented at the 1985 Annual Meeting of the American Statistical Association. He finds:

> Analytic studies of the size of the undocumented population indicate that it is significantly smaller than conjectural estimates [such as those of the *Denver Post* article] suggest. The studies reviewed here and the synthesis involving new data point very strongly to a range of 2.5 to 3.5 million undocumented aliens as the best figure for 1980. [12, p. 18]

So much for the "20 Million Illegal Aliens."

2. *How Many Deaths from Artificial Food Additives?*

The director of the Center for Science in the Public Interest, Michael Jacobson, was reported to have said: "I'd estimate that a maximum of 10,000 to 20,000 deaths per year could be attributed to artificial food additives [13]."

Dr. Melvin Bernarde, of the Department of Community Medicine and Environmental Health at the Hahnemann Medical College and Hospital of Philadelphia, felt that if that many people died each year due to cancer-causing food additives, he would call it an epidemic and said:

> Few are going to question where or how Jacobson obtained his figure of 10,000 to 20,000 deaths per year. It will be accepted as fact—because it's in print. I've tried to corroborate his figures but can't. Not because no one will give me the data, but because no one appears to have them. Yet they are, thanks to Jacobson, now part of the public record—to be quoted and requoted. [14]

Once again, the attention-getting number may be high, low, or even fairly accurate. But without a source, we can't make our own evaluation and the number is worthless for making a decision about personal or public policy in regard to food additives.

3. *How Long Do You Think You Will Live?*

In a laudatory article about a nutritionist from New Zealand, Dr. Michael Colgan, who has developed an "individualized vitamin-and-mineral program," we find the following statement:

> The urgency of Colgan's mission—and it can be called that—is most clearly understood against the backdrop of nutritional statistics that dwarf our attempts to "eat right and stay healthy." Here's a small sampling:
>
> In both the United States and the United Kingdom the average life expectancy of an adult, twenty-five, has not changed for more than 30 years. [15]

There are no U.S. or U.K. government sources for this statement given in the article. However, we can easily check this statement for the United States. The official figures show that the average number of years of life lived beyond age 25 ("life expectancy at age 25") for the total U.S. population increased from 46.6 years in 1949/1951 to 50.7 in 1980; an increase of 4.1 years. Life expectancy at age 25 increased for both men and women and for both whites and nonwhites [16, p. 8; 17, p. 74]. The author of the article not only gives no source, but the figures in the statement quoted above are wrong. In view of this, how much credence can you give to the other statements and nutritional recommendations in the article?

We encourage skepticism, but we do not want to encourage unfairness. We don't know what Dr. Colgan and the nutritionists quoted in the article actually *said*, we only know what the author of the article *reported*. Thus, without additional knowledge, our skepticism is directed at the reportage and not at the subjects themselves.

4. I've Got My Sources

The government of the USSR took a census in 1979, but did not publish any information on the age distribution of its people.* In the absence of this information, demographers outside the USSR try to estimate this age distribution. Since the USSR published information from their 1959 and 1970 censuses, demographers can make estimates for 1979 by using assumptions based on those two sets of published data. Each demographer, however, obtains slightly different numbers, since each makes somewhat different assumptions.

Murray Feshbach recently published what he implied is the "true" age distribution—not an estimate—for the USSR in 1979. Where did he get his information? How did he get it? Where did it originate? All he gives us is a statement about "New population data I have recently acquired," and: "I fully believe that these data reflect the official age distribution from the 1979 census of population and am publishing them under my own cognizance" [18]. He gives no further information as to how or where he obtained these "new population data." He does not compare them to estimates by other researchers. He does not tell us his sources, so for all we know, his numbers may be no more than another estimate. Are we to believe that Feshbach's data are in some way uniquely correct? It is the essence of the scientific method that we should be able to verify his conclusions.

The U.S. Census Bureau has confounded this issue by giving Feshbach's figures an "official U.S." imprint in a paper by W. Ward Kingkade [19]. The bureau is scrupulously careful in describing how it collects the U.S. census data and evaluates its accuracy. Why, then, did it accept Feshbach's unverified numbers? Is this another set of mythical numbers in the making?

III. Wrong Numbers

A. Miscounting

How easy it is to collect or record numbers incorrectly! In any complex data collection, many people are involved. Not all "get the word" and not all are capable of carrying out the instructions. It is not unusual for data collection and recording to take place in a work environment where the people are required to carry out several tasks at once, and data may get mangled and abused.

*We discuss the withholding of data by the USSR and other governments in Chapter 14.

1. *Policing the Police*

Arrest statistics are crime's basic data, and producing these statistics dramatically illustrates the difficulties of data collection and recording. Considerable ingenuity and effort have been applied to the problem of arrest statistics. Unfortunately, ambiguity in the definition of an arrest has been a continuing problem. After a long history of judicial and statistical concern for just what is an "arrest," arrest counting rules have been established and widely disseminated through training as a part of the Uniform Crime Reporting (UCR) program. The rules are simple enough. For example, "Count one (arrest) for each person (no matter how many offenses he is charged with) on each separate occasion (day) he is arrested or charged" [20, as quoted in 21].

Skeptics have suggested that police departments increase or decrease the reported number of arrests in response to political pressure. The argument is made, and supported in some cases, that the greater the number of arrests, the better the police agency is doing its job.

But a study by Sherman and Glick for the Police Foundation shows that police departments underreport as well as overreport the number of arrests [21, p. 27]. The major decision affected by arrest statistics is which policy to use to deter crime: aggressive police patrol practices or more arrests. Sherman and Glick give an excellent summary of the problems of getting correct numbers for arrests; their observations have application to many fields.

Glick spent two weeks in each of four police departments, auditing the departmental counts for a month. He found counts of total arrests by department in error by −2%, +5%, −5%, and +14%. The error in count for all four departments combined was +1.5% out of 3584 actual arrests.

A sample of size four does not allow us to make meaningful projections to all police departments. But it does suggest that serious miscounts can exist. Can we take it for granted that individual errors in all the reporting police departments in the country will balance out in every reporting period? What are the consequences of local policing decisions based on count errors as great as 14%?

U.S. Representative Charles E. Schumer, a Democrat from Brooklyn, is concerned about the quality of another type of criminal justice data: the reports sent to the Federal Bureau of Investigation (FBI) by localities about suspects wanted for arrest. He called the poor quality data at state and local levels "a gaping hole" in the criminal justice system, saying: "The most effective way to improve society's ability to fight crime and to protect the lives of police officers and the civil liberties of all citizens is to make sure these records are accurate" [22].

The magnitude of the problem has been determined by the FBI's auditing force:

> David Mitchell, the head of the F.B.I.'s new auditing force. . . . said that the available evidence [from the FBI's recent audit] indicated that on a national basis

> approximately 6 percent of the 211,296 warrant entries and 4.5 percent of 165,253 stolen vehicle entries being transmitted through the [FBI's National Crime Information Center] system had serious flaws. [22]

Do we have to accept poor data as the price for having a nationwide, large-scale computer system? Not if we take some lessons from this example. At the start, the FBI

> ". . . felt that the accuracy of the information provided by the state and local jurisdictions was beyond our control other than repeatedly urging them to worry about data quality," Mr. Mitchell of the F.B.I. said. "But several years ago, our advisory board agreed we should do more than just lead people to water." [22]

It is just not good enough to encourage people to "worry" about the quality of their data. The quality of data should have been a primary concern when the system was first installed. However, the FBI's system was allowed to operate for 15 years before the FBI started its auditing program.

The computer era brings special problems. The FBI allows many agencies which are responsible to other authorities to enter data into the National Crime Information Center's computer. There is no verification process at the center's computer, and audits have shown that the quality level of data from different states varies dramatically. The data from one unnamed city was found to be of such poor quality that the FBI wiped out thousands of files that city had placed in the system.

None of these problems *must* occur. They can be prevented, and the FBI's auditing force's statistical work is the first step in identifying the problem, its sources, and potential solutions. Whatever means the FBI uses in the future to ensure the quality of its data can be used on a smaller scale by the individual states, localities, and agencies.

2. *Can You Count on Researchers?*

Even researchers are subject to miscounting. The research of Drs. Mark and Linda Sobell on alcoholism was challenged, both by subjects of the study and by Drs. Pendery and Maltzman, who claimed fraud. Since the issue was the important one of whether control of drinking habits or abstinence yielded better results, federally appointed committees in both the United States and Canada studied the Sobells' research. They were exonerated of any intent to deceive.

However, "The (U.S.) panel found that the Sobells made incorrect statements about the number of times they contacted patients . . ." [23]. Counting correctly isn't easy!

B. Misclassifying

When a scientist classifies an object for statistical purposes, other qualified scientists should classify it in the same way. We call this type of verification "the replicability of observational units." In some scientific fields (e.g., geology, physical anthropology, archeology) the failure of replicability is called "observational discrepancy." Paul Fish studied observational discrepancy in archeological classification and found "provocative and some surprising results" [sic] [24].

In one location, a single instructor trained experienced analysts in the taxonomic scheme for classifying one of the most clearly defined types of ceramic shards (pieces of pottery). Despite this, discrepancies in classification ranged from 22% to over 30% between any two analysts. The discrepancies were distributed almost evenly among the analysts and the types of items being classified. In some cases, analysts differed in the definitions of characteristics. In others, the same analyst changed definitions from one time to another.

Even when classification was based on characteristics measured using simple instruments, the analysts occasionally disagreed.

IV. Bad Measurement

A. One Way or Round Trip?

Years ago, one of us received a questionnaire asking "How many miles do you commute each day?" The question was confusing. Did it mean one way or both ways? There was considerable doubt in the minds of typical respondents, but it is not reasonable to assume that potential respondents are going to make a phone call (perhaps long distance) to ask the originator what was meant. We assume that most respondents made the assumption one way or the other and gave a number.

If the originator wanted the one-way commuting mileage, then the value estimated from this survey would be high, because some respondents answered with the round-trip mileage. If the originator wanted the round-trip mileage, then the estimated value would be low, because some of the respondents answered with the one-way mileage.

B. Will the Real Number Please Stand Up?

1. *The United States' Original Inhabitants*

How correct are the numbers appearing in the reports of the U.S. Census? Some may be as correct as the human mind can devise. Others must be taken with a degree of skepticism. We illustrate the application of skepticism to a particular case, the number of American Indians. The U.S. Census reported 524,000 in 1960, 793,000 in 1970, and 1,478,000 in 1980 [25, p. xi; 26, Table 74]. The reported numbers are given to the last digit as though they were correct to the last person, but we round off to thousands to avoid giving a spurious impression of precision.

How can we check these numbers? One way is to look at the percentage increase in population from decade to decade. This is easier with the American Indian population than with others, since this group increases in number almost solely by an excess of births over deaths. There is virtually no immigration of American Indians from foreign countries. A small number of Canadian Indians may cross into the United States and a small number of American Indians cross into Canada. No one knows how many make this transition, but in the absence of a movement large enough to be reported as a special event (which has not happened in the past three decades), the number can reasonably be assumed to be small.

The maximum possible increase in a population due to the excess of births over deaths during a decade is 30 to 35%. Between 1960 and 1970, the reported increase was 51%. Between 1970 and 1980, the increase was 86%. These are impossibly high rates of increase by the excess of births over deaths. One or another number is wrong—but which? At this time, we do not know. We can make several educated guesses. One is that the U.S. Bureau of the Census changed its procedures from one decennial census to the next. But we doubt that this can account for a large change, and it might decrease, rather than increase the reported population. Another possible source of discrepancy may be that people are changing their minds as to what ethnic group they belong to, since the U.S. Census relies on self-identification. Because of the increasing social acceptability of ethnicity in recent decades and an increase in concern for individual "roots," more people who have some Indian ancestry may be changing their identity to American Indian from one census to another. Or, the census takers may have failed to count some numbers of American Indians in 1960 and 1970. No one knows for certain and only a suitable survey holds the promise—but not the guarantee—of an answer.

To confuse matters further, the U.S. Bureau of Indian Affairs, another source of information on the number of American Indians, reported about 700,000 "official Indians" in 1980 [27]. This number is about half that reported by the U.S. Census! Either number may appear in a media story, a politician's speech, or, alas, a research paper.

Do you find a discrepancy of a factor of two troubling? Consider the answer to the question: How many people lived in the Western Hemisphere when Columbus arrived in 1492? The estimates vary from about 8 million to about 100 million [28, p. 661 ff.].

2. The U.S. Mainland's Newest Inhabitants

Another round in the "number, number, who's got the number" game resulted from the migration of Puerto Rican population to New York after World War II. As Clarence Senior described the situation: "With the end of World War II, plane loads of Puerto Ricans began arriving in New York City in response to the employment opportunities which the city offered . . . and they are facing the same social and economic problems as the earlier immigrants'' [29, pp. 1–2].

New immigrants always seem to pose problems for the receiving communities and the situation with regard to the Puerto Ricans was no different. Thus, both official New York sources and the media greatly exaggerated their numbers during the late 1940s: ''Prior to 1948 . . . various estimates placed the Puerto Rican population of New York City at half a million and upwards. In 1948 a Columbia University study [by Senior] showed there were not more than 200,000 Puerto Ricans in New York City. . . . if that many [29].''

The U.S. Census for 1950 reported 245,000 Puerto Ricans in New York City. In the two years between the Columbia University estimate and the Census of 1950, some 50,000 migrants from Puerto Rico to the mainland were reported; the Columbia University study's estimate was verified by these official numbers. Would the average newspaper reader have been aware of this exaggeration in newspaper stories which doubled the actual number of migrants [29]?

C. Even in the Best of Circumstances . . .

The preceding examples pertain to the measuring instrument—to making sure that the data collection does not introduce flaws in the data available to the investigator. But we also have problems at the respondent's end. Even with the best of intentions and the clearest definitions of the required measurement, bad data can result.

In a statistics course that one of us teaches to Master of Business Administration (MBA) students, there is a class exercise designed to illustrate data collection and analysis. The basic data are the students' incomes. The exercise is as follows: The instructor suggests that the class might be interested in the income of the group as a whole. The class is allowed to discuss ''income'' in depth, raising such questions as: ''Why do we want to know?''; ''What kind of income do we want to deal with (i.e., salary, salary plus bonus, salary plus outside income from investments)?''; and so forth. The discussion continues until the class reaches consensus. In the ten years this exercise has been offered, the decision is almost always to collect data on salary only and to prorate part-time income to a full-time equivalent.

With the general goal settled, the discussion turns to how to collect the data. Usually, the class decides to collect the ''annual salary'' (actual or prorated) as the basic data. The class must then decide on the exact format of the data to be collected. Should it be expressed as annual salary in dollars? With or without comma for thousands? Annual salary in thousands? With or without a ''K'' to indicate thousands? Leading zeros? Dollar signs? How are the unemployed to indicate their situation? Is there a special indication for part-time workers?

The exact format of the response is determined by consensus. Then index cards are distributed and the class members put their responses on them in accordance with the rules just determined. The instructor collects the cards and the responses are tabulated. The average class size is 30 and in over 10 years, there has never been a class in which fewer than three responses are defective!

Some respondents will give their annual salary in dollars (with and without commas), or give their weekly salary in dollars or thousands, or spell out the value, or give two values (always in violation of the rules just agreed to), and so forth.

What is really remarkable here is that this is the result obtained when the respondent is also the designer and data collector, the definition of what is to be collected has been discussed for more than 30 minutes, there is no pressure on the respondent, the respondent has been able to raise any issues of ambiguity and have them resolved, and the respondents are graduate students with business experience. Imagine what kinds of data are collected when the question is asked with little, if any, explanation and possibly in syntactically incorrect form, or when the respondent perhaps isn't interested in the survey. That is the situation in too many cases.

Even when the question is clearcut, explicit, well-stated, and all reasonable precautions made to avoid erroneous response, there are questions for which answers cannot be accurate. Much so-called market research uses questions such as: "How many pairs of socks did you buy this year?"; "How many times has your dishwasher been repaired since you purchased it?"; "What is the average amount you spend on a restaurant meal?"; and "How much money do you have in your checking account?" Setting aside all questions of confusion about the nature of the answer, bias, or desire for secrecy, most people do not have the mental storage and recall capacity to answer these questions. How close to garbage must be the answers given to such questions, which then are analyzed at great length using complex statistical procedures and ultimately offered to the public as "real" information.

D. Underdeveloped Birthrates in Underdeveloped Countries

As we have discussed, disagreement among several statistics purporting to measure the identical concept immediately leads us to suspect that the underlying data are of low quality. Such discrepancies often are clues to the fact that the reported numbers are guesses and not estimates. We illustrate this point with statistics purporting to show changes in the birthrates of developing nations, an issue of considerable political and social importance. In this area, the authors of an article published in the *New York Times* express no doubts. They find that "Survey Reports Fertility Levels Plummet in Developing Nations" and make no mention of conflicts in statistics on birthrate [30].

Table 5.2 shows some of the published birthrates for three Third World countries, Brazil, Nicaragua, and Egypt, as assembled and analyzed by Joseph Cavanaugh [31]. We can calculate high and low limits to the estimates of changes in birthrate between 1965 and 1974 using combinations of the extreme values of the birthrate estimates. For example, for Brazil, we could estimate the birthrate in 1974 as being 80% of that in 1965 by dividing the low 1974 birthrate estimate (36) by the high 1965 birthrate estimate (45). Or, we could estimate the birthrate in 1974 as being 103% of that in 1965 by dividing the high 1974 birthrate estimate (38) by the low 1965 birthrate (39) estimate. Thus the birthrate change estimates for Brazil could range from a decrease of 20% to an increase of 3%. Table 5.3 shows the results for these three countries.

What is the "correct" change in the birthrate? Joseph Cavanaugh showed similar figures for 30 countries and the variability and conflict illustrated above were present in all these countries.

Table 5.2 Birthrate per 1000 of Population

		Estimate		Source	
Country	Year	High	Low	High	Low
Brazil	1965	45	38	d	e
Brazil	1974	39	36	c	d
Nicaragua	1965	49	44	e	a
Nicaragua	1974	48.3	44.5	a	b
Egypt	1965	42.5	41.1	f	d
Egypt	1974	38	35	e	c

Primary sources: (a) United Nations Statistical Office; (b) U.S. Bureau of the Census; (c) the Population Council; (d) United Nations Population Division; (e) USAID Office of Population; (f) the Population Reference Bureau.
Secondary source: Reference 31, Table 2, p. 288.

Table 5.3 The Range of Estimates of Percent Change in Birthrate, 1965 to 1974

Country	Range
Brazil	20% *decrease* to 3% *increase*
Nicaragua	9% *decrease* to 10% *increase*
Egypt	18% *decrease* to 8% *decrease*

It is a misuse of statistics to use whichever set of statistics suits the purpose at hand, and ignore the conflicting sets and the implications of the conflicts. We have great ignorance of birthrates in Third World countries, and this should surprise no one. If you look at the difficulties we have experienced enumerating the large and diverse population in our own country, what can we expect in less organized societies? How many births take place daily in Third World countries for which no record exists?

Did the originator of the report or the writer of the newspaper story have a secret wish (conscious or unconscious) that the birthrates should go down? Or just bad data?

V. The Unknowable

Some data can never be obtained. For example, there is a great deal of interest in the "underground economy" of the United States because the monetary transactions in this economic sector are not recorded and cannot, therefore, be used to verify that the correct taxes have been paid. This lack of information concerning transactions also concerns economists because it affects their ability to estimate economic variables that are important to economic decision making.

Some economists argue that one way of estimating the amount of money moving in the underground economy is to observe the flow of money into financial institutions [32]. This only represents, at best, the money *saved* in financial institutions. Since underground *spending* cannot be estimated, it is difficult to estimate the savings rate for the nation as a whole with this one important sector unknown.

VI. Disappearing Data

The best way to get statistical evidence for cause-and-effect relationships in human behavior, in most cases, is to follow specific individuals over a suitably long period of time. This is done by reinterviewing (or reexamining) them at intervals (day, month, year) for the period of the study. We call this a "longitudinal" survey. Without a longitudinal survey, data on the past (feelings, actions, and so forth) can only be obtained from present memories. Such memories are clouded by the deficiencies in the individual human memory and selectively recalled and modified according to current interpretations.

A. Getting the True Dope

Drug use (and abuse) is a major social problem in the world. There is a great need for public and private policy and direction concerning cause, cure, and control. To be effective, such policy should be based on a clear understanding of the underlying mechanisms. We seek such understanding from surveys as well as from experiments. But because of the ethical issues and the long-term nature of drug use and abuse phenomena, what we can learn from experiment is limited.

Many researchers have started (and some have completed) longitudinal studies of drug use and abuse [33–35]. Valuable information has been gained, but:

> A major problem [is] . . . the differences between those persons who complete questionnaires or are interviewed in all waves of a panel study and those who are captured in the first wave but are lost in later ones. . . . Of those seen in the first wave, only a percentage are seen in the later ones. It is usually only those who furnish data in all waves on whom data analysis is based, but if a large percentage of the original random sample is lost the generality of the conclusions becomes questionable. [34]

Just how many of the original starters were lost? For three studies [33, as quoted in 34] the proportions of original starters present in the final wave were 73%, 66%, and 44%.

Do we have reason to believe that dropouts from these longitudinal studies were random occurrences and therefore do not affect the conclusions? Alas, no. For "it was precisely the drug users, poorer students, and truants who were lost. . . . Data on . . . [other] percentages lost . . . also indicate that it is lower-class, minority-group members who are most likely to be lost" [34].

This does not mean that the data on the survivors to the final wave is invalid. What it does mean is that the data for them cannot be generalized to the original target population as a whole, that some statistical methods have little validity, and that, since data are missing for the most affected groups, the research may not fulfill the original purpose of the survey.

B. Jobs for Youth

Another study was concerned with the possible effect of initial job experiences on subsequent employment: "The premise . . . is that in the early career of a young person's life initial job experiences and attitudes are critical in shaping ultimate unemployment experience" [36, p. 2]. To test this premise, the investigator analyzed the age 14–24 cohorts of the National Longitudinal Survey, following males from 1966 to 1975 and females from 1968 to 1975.

Losses from the initial sample were considerable. As well as we can estimate from the information given [36, p. 37], about 25% of the male respondents were lost in the first three years. It is impossible to determine how many disappeared from the study in the following six years. No loss information is available for females. This does not mean that the researchers do not know how many subjects were lost, only that in their published paper they do not explicitly tell us how many were lost and do not give us a basis for calculating how many were lost.

In this case, we don't know: (1) the proportion of male and female survivors to the end of the study; and (2) their characteristics (as we know in the case of the drug survey just discussed). Thus, we have no basis for judging the applicability of the results. The author carried out considerable data manipulation, but no amount of computation can make up for the deficiencies in the basic data.

VII. Summary

A. Our Recommendations

The single greatest lesson you can take from these examples is this: no matter what your role, be it writer, reader, researcher, student, or teacher, you must always be skeptical.

For the reader:

If data or results based on data are given, look for a source in the article or report. If you find no source, look for one. If you can't find one, try to get information from the author. If the author's source is an article or report, look it up.

If you find no source and cannot get one from the author, look for data which will confirm or discredit the data. If your best efforts lead you to believe the data are mythical, unknowable, or unknown, then draw your own conclusions accordingly.

For authors:

Give sources when you give data or results of analyses. Don't base headlines and conclusions on mythical, unknowable, or unknown data. Help the reader by giving enough information about the data collection process to enable the reader to judge the results.

B. Are We Asking Too Much?

We don't believe that we are asking too much of either readers or authors. Not everyone agrees with us however; the following is a criticism of the preceding recommendations for authors from a statistician:

> It should be recognized that newspaper articles exist under restrictions of space. Also, while I don't condone the way the media often distort reports to make them interesting, I think it's obvious that if they surrounded everything with scholarly heaps of source references and discussions of procedures they would lose most of their readers. [37]

We respectfully wish to differ. As we note several times in this book, when the *New York Times* reports on a sample survey, it gives a description of the sampling methodology, the magnitude of the sampling error, and a brief discussion of the meaning of the relevant technical terms. When *Redbook* reported the results of its survey of sexual attitudes, the editors made space for a caution about the self-selected nature of the sample [38].

It is easy to use statistics correctly when you want to! In a letter to the editor in the *New York Times* (a type of newspaper item in which space is at a great premium), the secretary-treasurer of the American Philological Association, Roger Bagnall, finds it possible to be thorough about naming sources. We excerpt the relevant paragraph of his letter and italicize selectively to show how you can meet our standards for avoiding the misuse of statistics that comes from not giving source information:

> The output of new Ph.D.s in classics has fallen by about half in the past decade, *according to the National Research Council: from 88 in 1974 and 93 in 1975 to just 44 last year*. By contrast, the number of jobs advertised through the *American Philological Association's placement service has increased in the past few years to last year's (1984–85) 98 firm and 29 possible positions*. Not all of these are tenure-track, but 40 of the positions advertised last year were tenure-track *according to the copy of the advertisement*. [39]

It can be done, if the will is there. Do it!

6
Graphics and Presentation

Look here, upon this picture, and on this.

Shakespeare

And shall shew signs and wonders, that they may lead astray, if possible, the elect.

Matthew 24:24

I. Introduction

To the ancient Chinese proverb "one picture is worth 10,000 words," we add "for good or for bad." We all know that a dramatic graph can have a far greater impact than any somber table of numbers. John Tukey, the famous statistician, says that graphs enable us to "notice what we should have seen in the first place." Graphs are excellent for presenting some kinds of results in certain circumstances, especially to nonstatistical audiences. But the choice of presentation method is complex. For a summary of current research findings which includes the role and place of graphical presentation, see Reference 1.

Dr. Nowall, our fictitious researcher, wisely includes graphs in his reports and studies wherever and whenever he can. Dr. Nowall has discovered that threats of disaster or threatening changes based on statistics get him publicity on the front pages of the newspapers and invitations to appear on TV shows. He has learned that no word is as dearly beloved of the headline writer as "soaring." He always tries to find a way to plot his data to give the impression of a "soar." Even if the threatening quantity is really declining, he may find a way to plot it so that it soars on the graph, striking fear into the hearts of the readers.

Sitting in front of his personal computing workstation, he can create myriad graphs from one set of data. He can vary a graph to produce the effect he desires, which not infrequently shows some number rising astronomically. Of course, the use of the computer only extends his ability to manipulate graphs to suit his purposes, for even when he had only pen and pencil he was pretty good at this.

Edward Tufte and Howard Wainer have given professional statisticians and nonstatisticians many new insights into this process [2, 3]. We find that Dr. K. Nowall has much to add to our knowledge of these misuses. We look at some of the good doctor's past and present work, and also at several actual cases, in the hope that by doing so we will reduce the effect and number of graphical misuses in the future.

II. Some Horror Pictures

A. The Great Land Boom in the Suburbs of Los Angeles

The lead paragraph of the article called it "the biggest land boom in American History" [4]. To show the dramatic growth in land values, the article included the graph shown in Figure 6.1. The growth in values from April 1970 to October 1977 appears to be truly outstanding. At first glance, it looks as though there has been an increase of about forty-fold during this period, since the ratio of the height of the line plotted for October 1977 is about 40 times that of the height plotted for April 1970. This effect occurs because all the values below the real estate index value of 140 are deleted, greatly exaggerating the apparent ratio in values. When the graph is redrawn (Figure 6.2) with the left-hand scale carried down to zero, you can see that the true ratio of growth is about 2.6, not 40. Not a bad gain in land values, but not as dramatic as the published graph with its missing zero on the vertical scale.

It is a major statistical sin to draw a graph with the vertical (left-hand) scale cut short so that it does not go down to zero. We call this the *sin of the missing zero*. This major sin creates misleading impressions. Unfortunately, many graphs are drawn this way.

Figure 6.1 is an example of a lesser sin as well. Successive numbers in many economic time series often come from a special pattern of growth. For example, increases in wages and prices are usually based on percentages, rather than values. How are the demands of workers such as those involved in construction of homes in Orange County, California, usually stated? In terms of a percentage increase each year. In multiyear contractual negotiation, you will often see settlements for raises such as "5% per year for three years."

Many prices in an inflating economy are increased by percentage; price

ORANGE COUNTY HOME PRICES
April 1970-October 1977

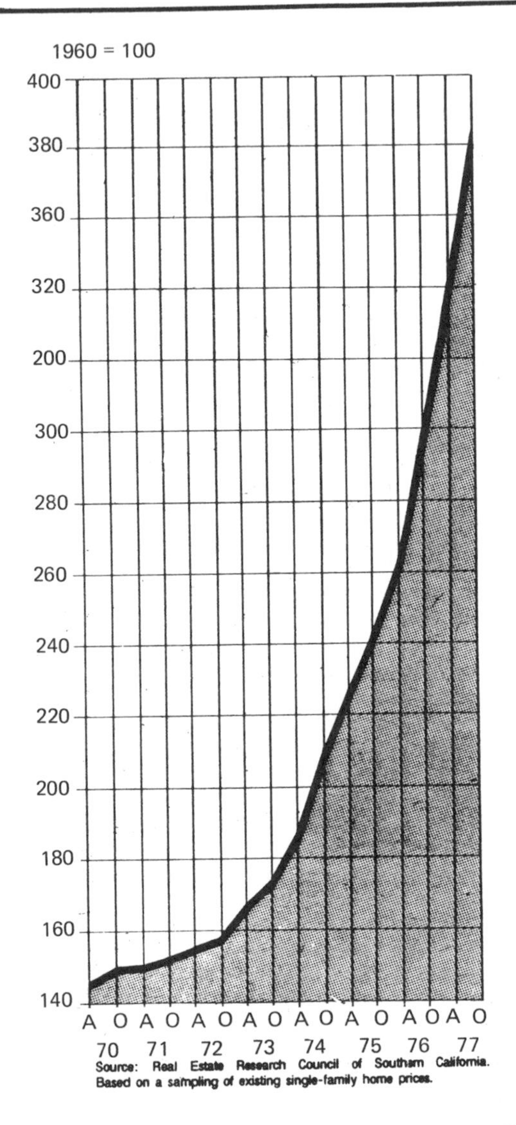

Figure 6.1 The "soaring" home prices in Orange County. Soaring is emphasized by the *sin of the vertical axis*, no return to zero or indication of the absence of zero on the vertical axis. The designations "A" and "O" along the horizontal axis denote "April" and "October," respectively. (From Reference 4, reprinted by permission of *Barron's*, copyright 1978 by Dow-Jones & Company, Inc. All rights reserved.)

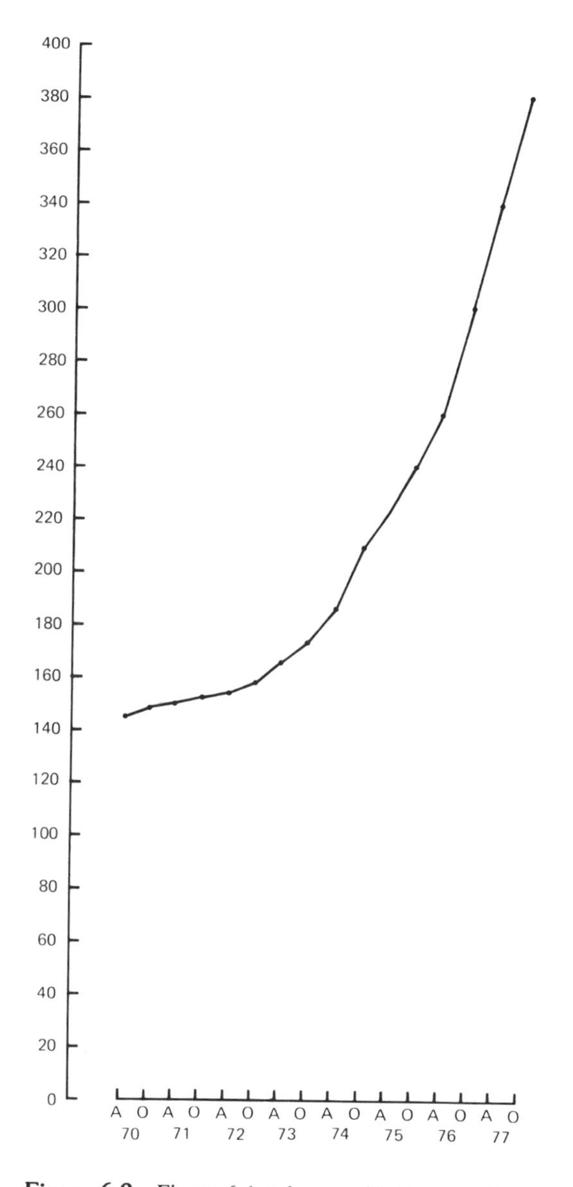

Figure 6.2 Figure 6.1 redrawn with the zero shown on the vertical axis. The magnitude of soaring is clear because you can now see the value of the index in April 1970.

increases are announced sometimes as "all prices in our catalog are increased by 10%." Wage increases are almost invariably based on some percentage. For this reason, inflation is also measured as a percentage increase per year. The most common way to measure the effect of inflation is to compute the *compound interest rate*, which is often compared to the compound interest paid by certain investments, such as bank savings accounts, certificates of deposit, etc. If you put $20,000 in a bank savings account paying 10% interest compounded annually—and leave both the $20,000 principal and accumulated interest in the account—your investment will grow to $22,000 at the end of the first year ($20,000 plus 10% of $20,000). At the end of the second year, it will have grown to $24,200 ($22,000 plus 10% of $22,000), and so on. At the end of five years, your investment will have grown to $32,210, a ratio of 1.61.

We use this reasoning to evaluate the rise in home prices in Orange County. The real estate value index shown in Figure 6.1 grew from 145 in April 1970 to 380 in October 1977, a ratio of 2.62. To measure this growth by a compound interest rate, we ask this question: What is the constant annual compound interest rate that will cause a growth of 2.62 in 7.5 years, the growth ratio of the real estate value index? You can look up the answer in an interest table, available in most banks or in books on finance, or use a calculator or computer.* With any one of these methods, you will find that the equivalent compound interest rate for the growth in this real estate value index is about 14%.

How does 14% compare to the inflation rate during the same period? The average inflationary increase in the Consumer Price Index for All Urban Consumers [5] from April 1970 to October 1977 corresponded to growth at a constant percentage increase of about 7%.† Figure 6.3 shows the "soaring" graph produced by plotting a constant 7% annual increase from April 1970 to October 1977.

So what's the story? During this period, real estate in Orange County was growing at an average rate of 14%, about double the rate of inflation. Not bad, but not as soaring and booming as the casual observer of Figure 6.1 might have thought.

Let's now look at how these changes can be shown properly on a chart. Figure 6.1 was drawn to an "arithmetic scale." The vertical scale is "arithme-

*The basic relationship is $(1 + i)^{7.5} = 2.62$, where i is the constant annual interest rate that would give the observed growth of 2.62 in 7.5 years, the period over which the growth occurred. The interest rate is expressed as a proportion, where $i = .1$ for a rate of 10%. You want the solution to $i = 2.62^{1/7.5} - 1$, which you can find on many pocket calculators. The 7.5th root of 2.62 is 1.14; thus $i = .14$, and the equivalent compound interest rate is 14%.

†There are several indexes of inflation or price level changes, but for all practical purposes they give similar results.

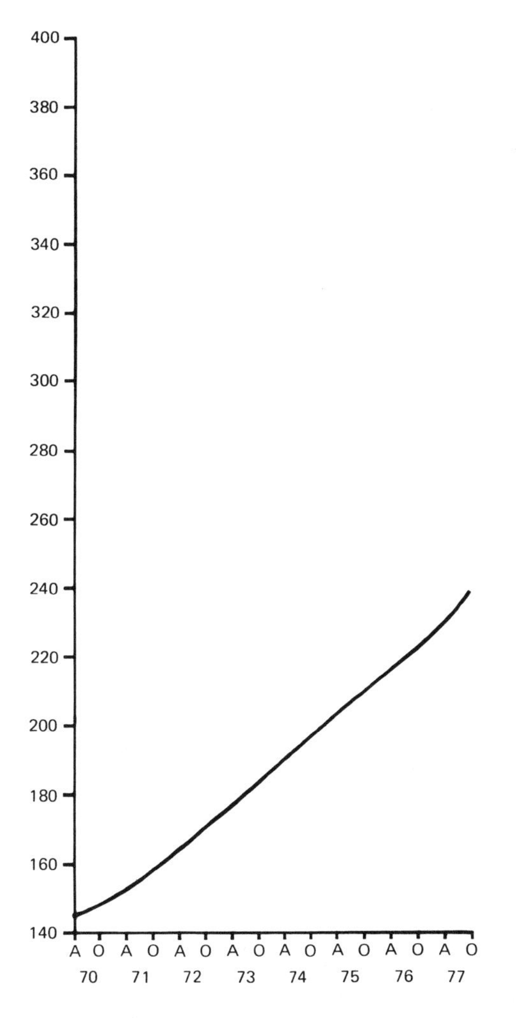

Figure 6.3 This is the graph of a 7% per year growth during the period April 1970 to October 1977 plotted to an arithmetic scale on the vertical axis. See how it "soars."

tic," which means that an increase (not a percentage, but a constant value) in the index covers the same distance along this axis no matter where it occurs. Thus, the plot of an index that is increasing by 20 points each year is a upward-sloping straight line.

However, if an index that is increasing by 20% per year is plotted against an arithmetic vertical scale, it soars upward because the constant percentage increase covers a larger and larger distance along the vertical scale. A 20 *percent*

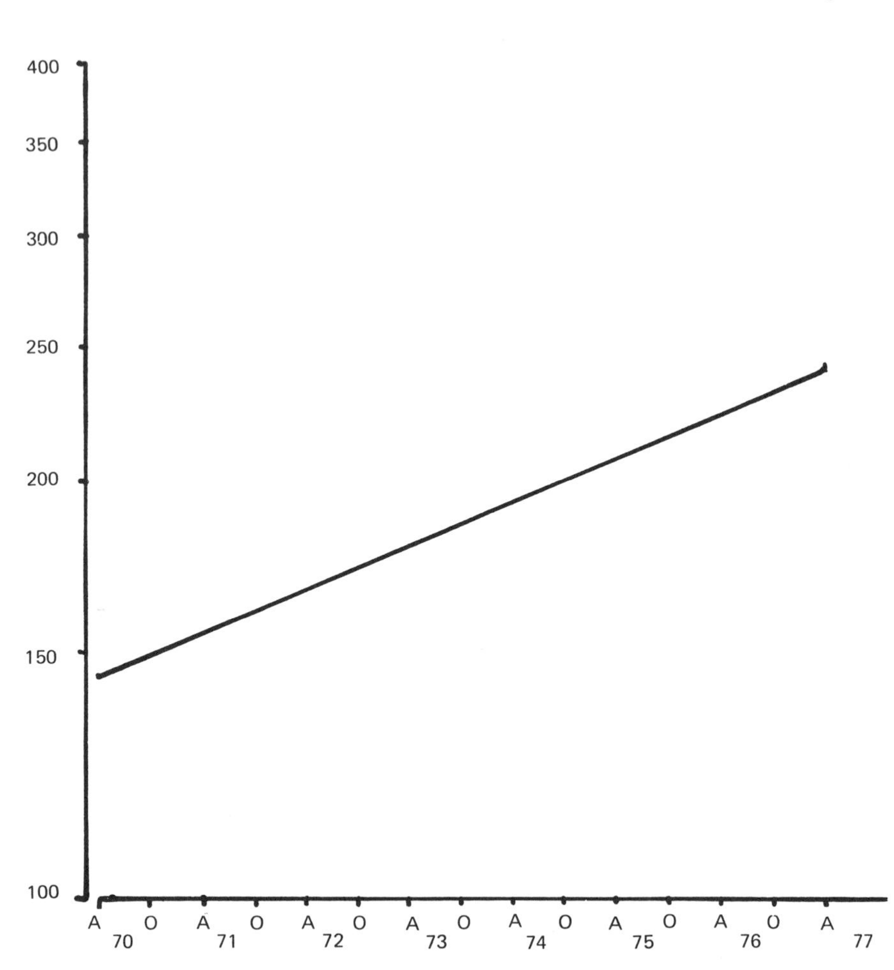

Figure 6.4 The graph of a 7% per year growth during the period April 1970 to October 1977 plotted to a ratio scale on the vertical axis. Note that the growth which appeared to soar in Figure 6.3 now corresponds to an upward-trending straight line.

increase when the index is 200 covers twice as much distance along this axis as the same percentage increase when the index is 100. But if you plot a time series against a ''ratio-scale'' vertical axis, the same *percentage* changes cover the same vertical distances.* The constant percentage change that ''soars'' on an arithmetic scale is a straight, upward-sloping line on a ratio scale graph.

*The vertical scale we describe is a logarithmic scale. When plotting a time series, a horizontal arithmetic scale is usually used and the complete graph is called a ''semi-log'' graph. You can get special preprinted semi-log graph paper in many sizes, or produce such a graph to your own specifications using a graphics computer program.

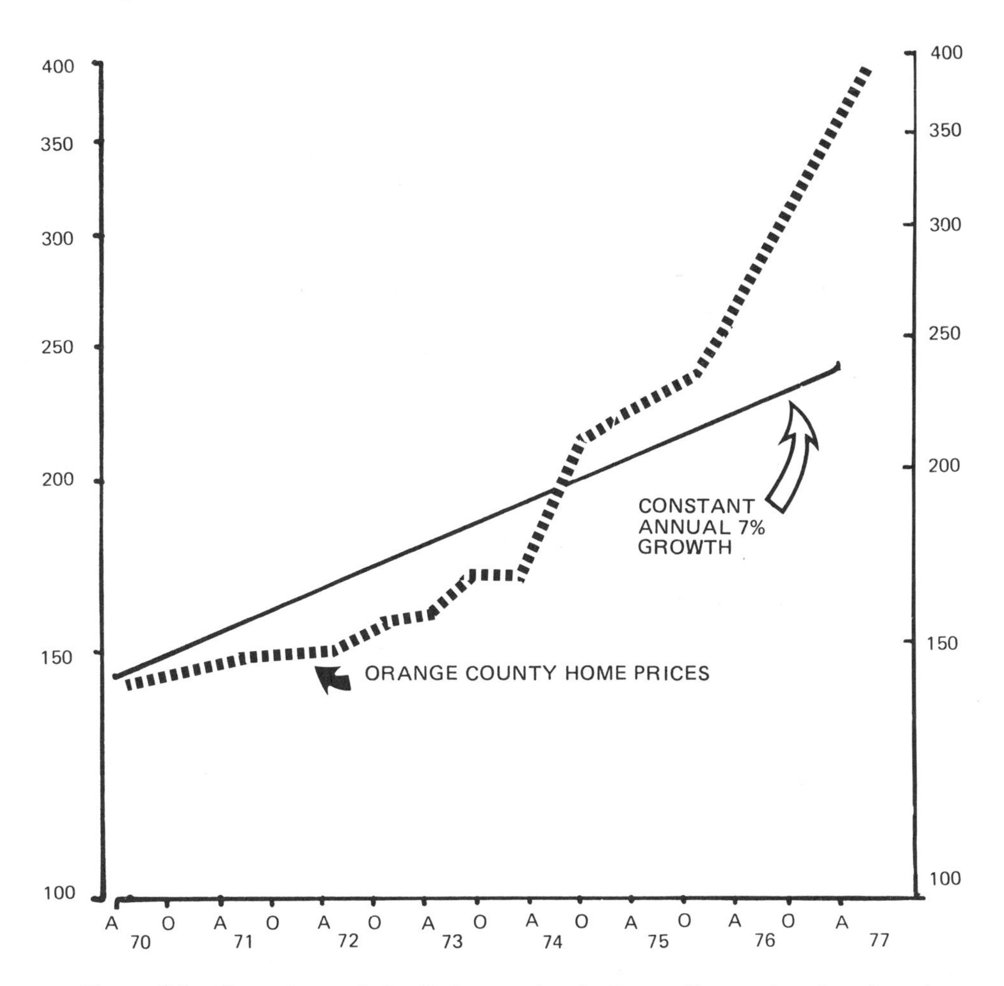

Figure 6.5 The real estate index for home prices in Orange County plotted on the ratio scale graph of Figure 6.4. The soar is now revealed as a nearly constant growth at an annual rate lower than the average inflation rate until 1974, when it climbs at a more rapid rate. This graph clearly shows that during the period of interest, the average annual percentage rate of increase of the index is twice that of the Consumer Price Index.

We now show how the use of a ratio-scale graph makes it easy to see—without the bias of a ''soar''—the changing real estate values in Orange County. First we plot the inflationary effect in Figure 6.4; in which a constant 7% annual percentage change in inflation is a straight line, because it is plotted on a ratio scale. (Note that ratio scales have no zero.)

To compare the growth of the real estate index during the same period, in Figure 6.5 we have plotted its values on the ratio-scale graph of Figure 6.4.

Clearly, the real estate index was increasing at a constant percentage rate less than the average rate of inflation until 1972, and after 1973 it increased at a nearly constant rate until 1977. In October 1977, the right-hand end of the plotted index line has risen twice as high from the start in April 1970, as the plotted line for 7% inflation. Thus, you can see clearly that, during this period, real estate prices in Orange County were growing at an average rate of 14%, about double the rate of inflation.

Comparing the rise in home prices with the overall rate of inflation is one way to look at these data. Another way is to compare the change in home prices to the change in family income. Homes are not paid for out of the inflationary rise in consumer prices, but out of family income. Unfortunately, we lack information on family income for Orange County for the years shown in Figure 6.1 and cannot make this comparison. However, when you see such a chart again, you should ask: What were the changes in earnings? In family income? In national income? It is important to compare the raw data with some measure of economic change that affects the home buyer.

B. The Japanese Are Coming: Sinning Against the Horizontal Axis

In the case of the great land boom (Section II.A above), we saw the sin of the missing zero on the vertical scale, in which the growth of real estate values was exaggerated. Less frequent, but just as sinful, is the sin of the distorted horizontal scale. In Figure 6.6, you can see the projected growth of Japan's multinational corporations distorted into a threatening ''soar'' by a manipulation of the horizontal scale [6]. On the original chart, the decade from 1970 to 1980 gets a horizontal distance of three times that given to the decade from 1980 to 1990. The effect magnifies the apparent ''soaring.'' To make sure that you get a strong soaring feeling, the graph's designer also exaggerated the effect by curving the horizontal scale upwards on the right side.

Look carefully and you will find that this graph misleads the viewer. During the first ten years (1970 to 1980) shown on the graph, Japan's ''overseas stake'' grew about 10 times. During the following ten years (1980 to 1990) Japan's ''overseas stake'' is projected to grow only 4.5 times. But the way the graph is drawn, this smaller growth appears to ''soar'' upwards. The average annual growth rate during the first decade is about 25% per year, and during the second ''soaring'' decade, the average growth rate is about 16% per year.

Thus, the three-fold compression of the last ten years (1980–1990) on the horizontal scale has made a lower growth rate (16%) loom larger than a higher rate (25%). A clear picture of the projection of Japan's overseas stakes is shown in Figure 6.7, which is a plot of the graph of Figure 6.6 on a consistent horizontal scale. We can only wonder whether the originator of this graph knew what he or she was doing. Can you find it in your heart to forgive this sinner?

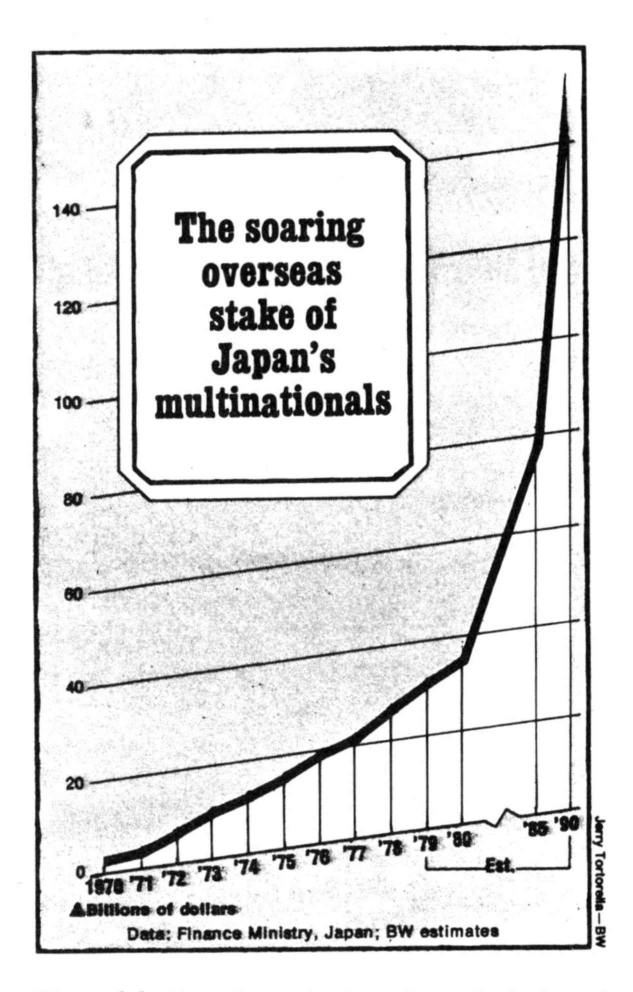

Figure 6.6 Two changes in the scale on the horizontal axis (1980–1985 and 1985–1990) give a "soar" to Japan's overseas stakes. Although one of the changes in scale is announced by a wiggle in the horizontal axis (between 1980 and 1985), the result is a serious misuse. (From Reference 6, reprinted from the June 16, 1980 issue of *Business Week* by special permission, copyright 1980 by McGraw-Hill, Inc.)

C. Double Out: Sinning Against Both the Horizontal *and* the Vertical Axis

Dr. Nowall was studying the United States birthrate for discoveries that would startle the world. But he had a problem that plagues many researchers: He could find no evidence of an effect, for there wasn't much variation in the birthrate (per 1000 of population) from 1973 to 1983. During that decade the birthrate in-

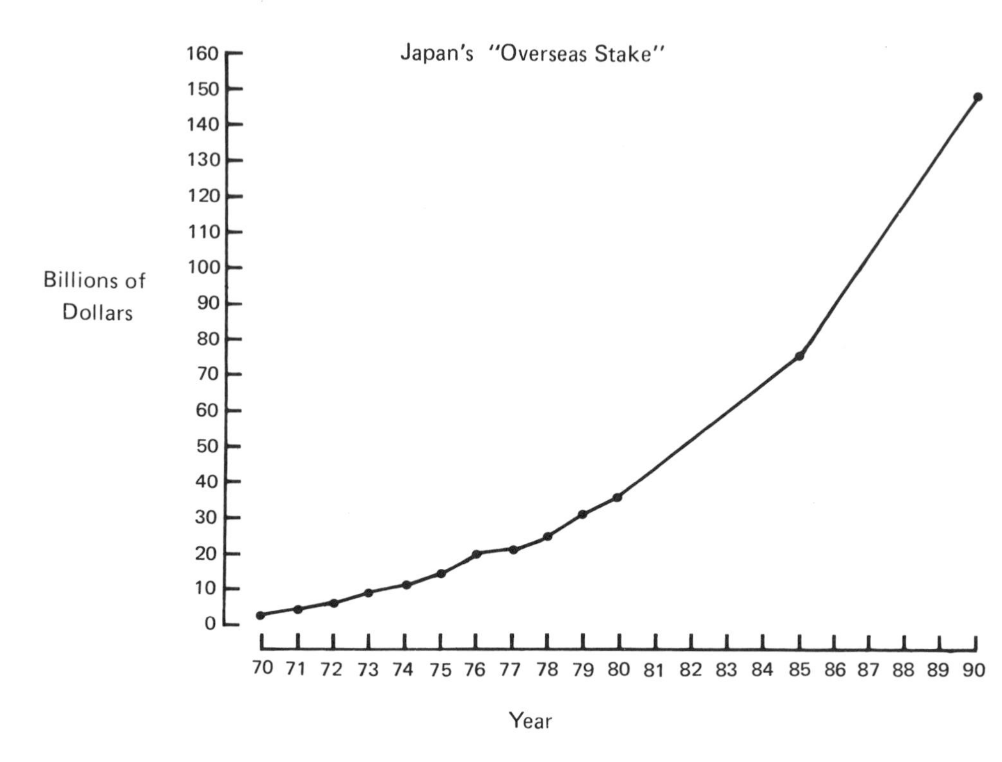

Figure 6.7 Figure 6.6 redrawn with a consistent scale on the horizontal axis. The resulting plot is still somewhat misleading. The *percentage* growth from 1970 to 1980 is almost double the projected percentage growth from 1980 to 1990, while the growth of Japan's overseas stakes is lessening.

creased by only about one birth per 1000 people in the United States. Alas, in an ordinary chart, this does not look exciting.

His fertile brain soon supplied a way to attract attention, as he mused, "If I make the vertical axis extra long and the horizontal axis extra short, I'll have an exciting and newsy story." The result is shown in Figure 6.8, which he headlined as "Birthrate Soars in Decade."

One of the editors who received Dr. Nowall's release recognized that Dr. Nowall had committed a sin against both axes. He redrew the graph as shown in Figure 6.9 and did not print an article based on this release. Note that neither Dr. Nowall (who had read about missing zeros in the *New York Statistician*) nor the editor committed the sin of the missing zero. It is an acceptable practice to break the vertical line to show that the vertical scale has been shortened.

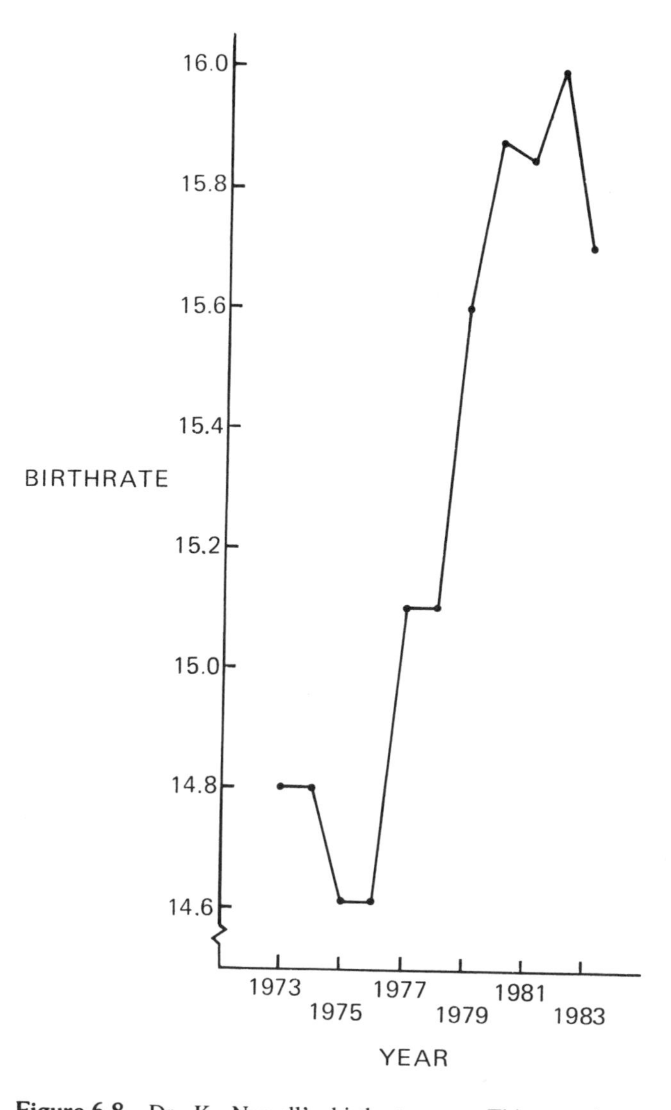

Figure 6.8 Dr. K. Nowall's birthrate soar. This soar is the result of a shortened horizontal axis, an expanded vertical axis, and not returning the vertical axis to zero. Note the use of the small "break" in the vertical axis to show that the axis is not returned to zero. This is a common practice, but is less satisfactory than a full break, as shown in Figure 6.9. (From Reference 8, Table 80.)

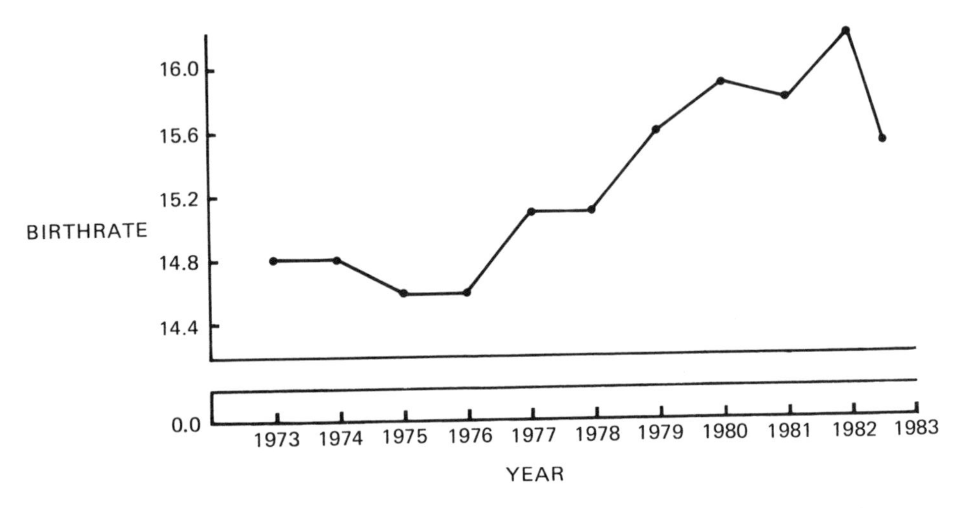

Figure 6.9 Figure 6.8 redrawn. The soar is reduced to a rise. This graph has a longer horizontal axis in accordance with Tufte's recommendation, "Graphics should tend toward the horizontal, greater in length than height [2, p. 186]." The break in the vertical axis to show that it does not return to zero is a "full" break for emphasis [14].

D. The Great Currency Mountain Is a Molehill

"Americans Hold Increasing Amounts in Cash Despite Inflation and Many Other Drawbacks," says the headline in the *Wall Street Journal.* Figure 6.10 shows how the currency pileup has been steepening in the past 25 years, as "the amount of currency in individual hands is soaring . . ." [7]. The rise from the years 1953 to 1979, from 28 billion dollars of currency in circulation to 102 billion, does seem dramatic, especially when you view the apparently increasing rate of rise at the right side of the graph.

But what were the real changes in the amounts of currency held by Americans? One way of measuring these changes is to reduce the dollar amounts by the amount of inflation, which gives us the change in the *purchasing power* of the currency in circulation. The ratio of the Consumer Price Index for 1979 to 1953 is 229.9 divided by 80.5, or 2.86 [5]. The ratio of currency in circulation for 1979 to 1953 is 3.6; the amount of currency in individual hands has increased by 26% more than is necessary to maintain a constant purchasing value. That is certainly an economic change of importance, but is the growth in currency in circulation as dramatic and as threatening when the purchasing value of currency in circulation is plotted as in Figure 6.11? In purchasing power, the currency in circulation is not "soaring"; it is "creeping" upwards.

Another way of measuring the significance of change in the amount of

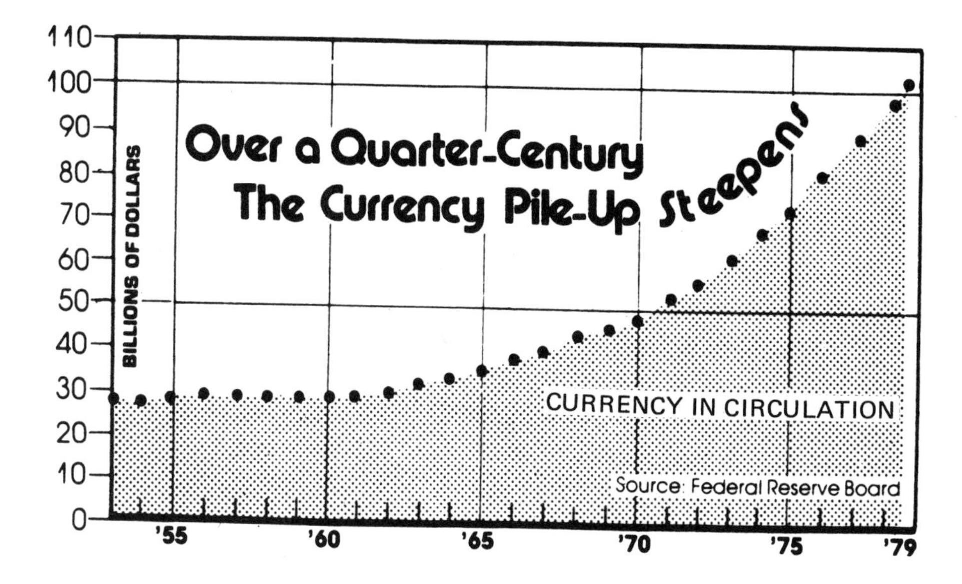

Figure 6.10 Another soar: The piling-up of currency in circulation. (From Reference 7, reprinted by permission of the *Wall Street Journal*, copyright 1979 by Dow-Jones & Company, Inc. All rights reserved.)

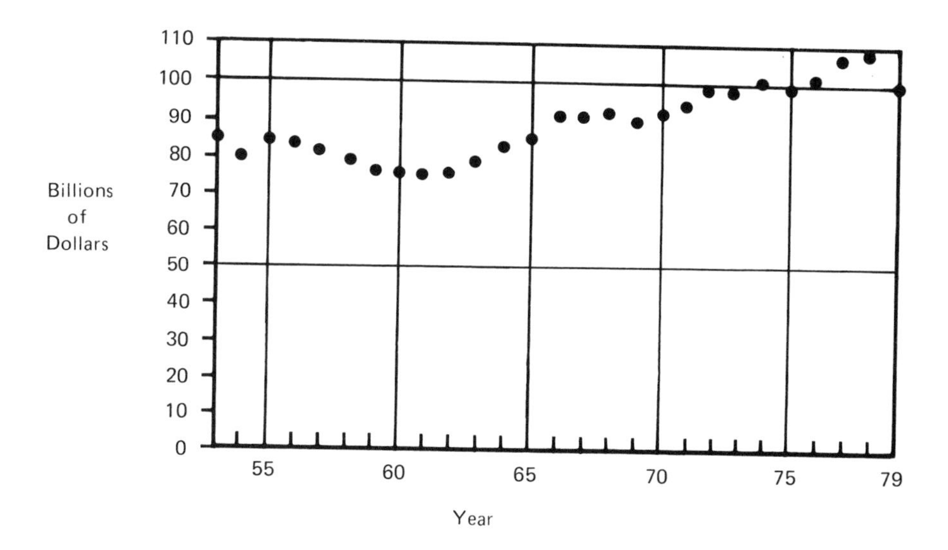

Figure 6.11 Figure 6.10 redrawn with the dollars in circulation "deflated" to account for the general inflationary trend.

currency in circulation is to compare it with changes in the total economic position of the U.S. population. We can use the amount of personal income—the current income received by persons from all sources minus their personal contributions for social insurance [8]—as a measure of total economic position. What we find is that personal income increased more rapidly than the amount of currency held by the public. From 1953 to 1979 the amount of personal income increased between six and seven times, but the amount of currency held by individuals increased only about four times. On a comparative basis, in 1953 the currency in circulation amounted to about 10% of personal income, but in 1979 it was only about 7%.

Increases in population can increase the amount of currency in circulation, so let us see if the increase in the population accounts for the observed increase in the amount of currency in circulation. If you divide the deflated dollars (85 billion in 1953 and 100 billion in 1979) by the population in those years (in 1953, about 160 million and in 1979, about 225 million people), you get about $530 of currency in circulation per capita in 1953 and about $450 per capita in 1979. This is consistent with the result, based on personal income, described in the previous paragraph.

Thus, the newspaper headline could just as easily have read "Americans Hold *Decreasing* Amount of Cash," based on Figure 6.12. You may feel the same way as Will Rogers, who said: "I hope we never live to see the day when a thing is as bad as some of our newspapers make it."

E. Remedial Care for Graphs About Health Care Costs

When Dr. Nowall talks of a "soar," we can expect to see a sweep upwards at the right side of the chart (as in Figures 6.1, 6.6, and 6.8). When a rise is less dramatic, but still headline-worthy if properly presented, it becomes an "escalation."

Thus we have "escalating health costs" accompanied by their escalation graph (Figure 6.13), obtained by the *Connecticut Business Journal* from a company supplying services to reduce per capita health care costs [9]. As you can see, Figure 6.13 shows an escalation of per capita health care costs with time. But what is the "escalation" in per capita health care costs as measured in purchasing power? Since the base for comparing these costs is 1965 (the year at which the chart starts), we "deflate" all subsequent costs to their equivalent in 1965 dollars using the Consumer Price Index [5], as shown in the graph of Figure 6.14. We cannot be sanguine about a doubling of the purchasing power of per capita health care costs, but how different is the effect produced on the reader by Figure 6.14 as compared to Figure 6.13!

This is an example of the sin of inflated dollars. Once again we have to ask:

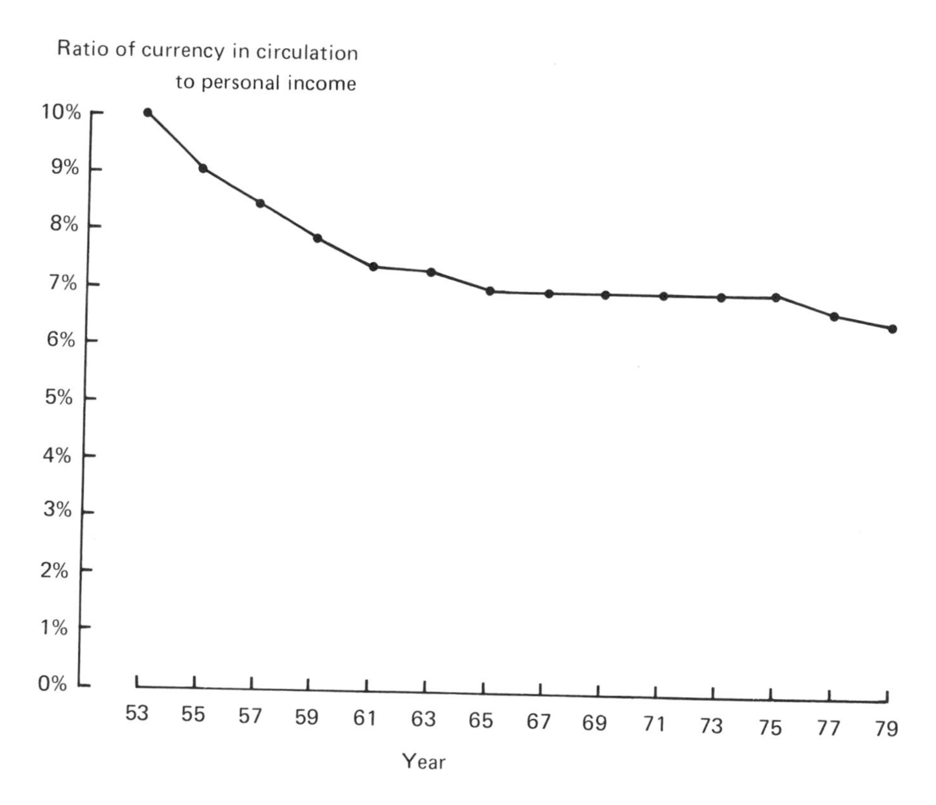

Figure 6.12 Is this the real story on currency in circulation? This is a plot of the ratio of currency in circulation to personal income.

Who benefited by committing this sin which gives an impression of a dramatic growth in health care costs?

F. More Action in the Graphical Horror Show

1. *When a Picture Isn't What It Seems to Be*

Real wages are wages corrected for the increase (or decrease) in the cost of living due to inflation (or deflation). For example, if the average factory wage rose from $20,000 to $22,000 in the next year (an increase of 10%), we do not know what happened to *real wages* without knowing the change in the cost of living. If, from one year to the next, the increase in cost of living was 10%, then real wages were not changed. If the cost of living increased by only 5%, then real

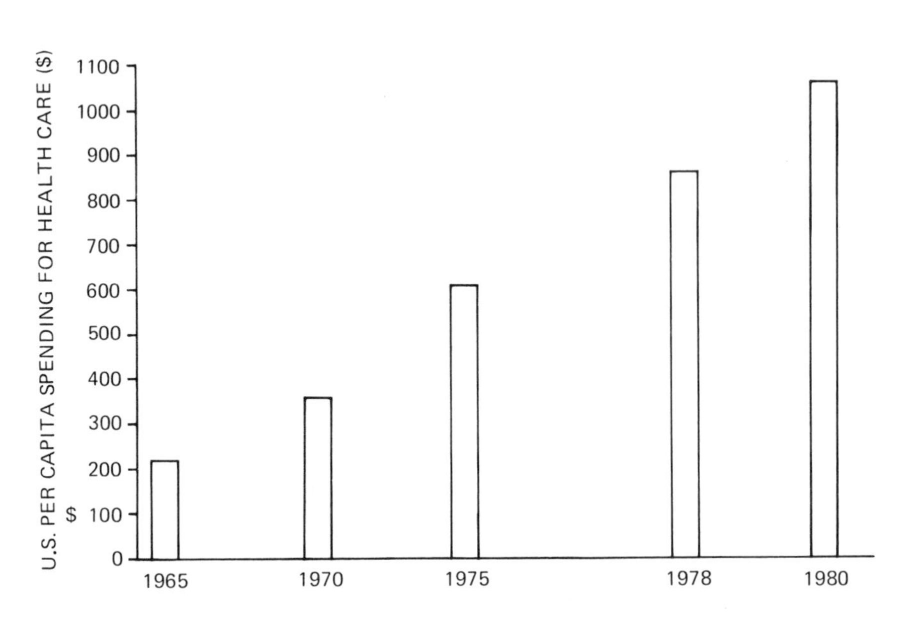

Figure 6.13 The shocking growth of per capita health care costs. The bar chart is plotted from the numerical values of per capita health care given in Reference 9 and to the same nonlinear horizontal scale. Note the unforgiveable sin against the horizontal axis. (Values of per capita health care from A. W. Hansen, Inc.)

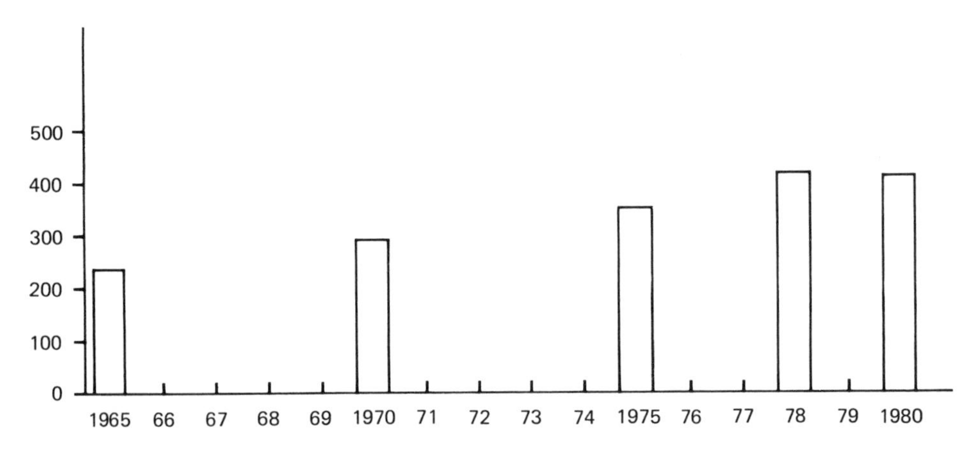

Figure 6.14 Figure 6.13 redrawn with the amounts "deflated" by the Consumer Price Index to account for the general inflationary trend. Deflation by other measures of price level gives essentially the same results. The horizontal scale has been made linear, which also reduces the apparent soaring of per capita health care costs.

Figure 6.15 Price levels rise by 3.4%, but this illustration shows how it feels, not what it is. (From Reference 10, copyright 1979 AFL-CIO News. Used by permission.)

wages *rose* about 5%; if the cost of living increased by 15%, then real wages *fell* about 4%.*

The pain of lost value in wages due to inflation is real, but its illustration in Figure 6.15 is false. The *AFL-CIO News* graphically reported the 1978 decline of 3.4% in real wages in Figure 6.15 [10], which the casual reader will see as a severe reduction. The decline is grossly emphasized by showing scissors cutting real wages at a point corresponding to a 40% decrease. This will certainly wake up the reader, but isn't this a sinful way to do it? This is known as the sin of false proportion. If the graph had illustrated the decline in true proportion (Figure 6.16), the impression would have been substantially different.

*The exact values are 4.76% (1.10 for wages divided by 1.05 for the cost of living) and 4.35% (1.10 for wages divided by 1.15 for the cost of living).

Figure 6.16 Figure 6.15 redrawn to true scale.

2. Just Plain Confusing

Now and then, we see a graph that is so poor as to be almost incomprehensible. Figure 6.17 is supposed to help us understand heat conservation in the home [11]. The title claims that it shows us sources of heat, in percentage. If that is all that is shown, why is the bar for the conventional home longer than the other three, which are all nearly, but not quite, the same length. What does the length of the bar mean, if anything? No clue is given. Does this chart help you to understand the distribution of sources of heat? Note that the labels are inside three of the bars, at the end of one bar, and below the bottom bar. This is a magnificent example of the sin of visual confusion.

The *Statistical Abstract of the United States* gives us some additional examples of variation on this theme. Figure 6.18 is a plot of Life Expectancy at Birth. This graph is confusing for several reasons: There are unnecessarily long and bewildering labels on some of the lines, and the visual conflict caused by the

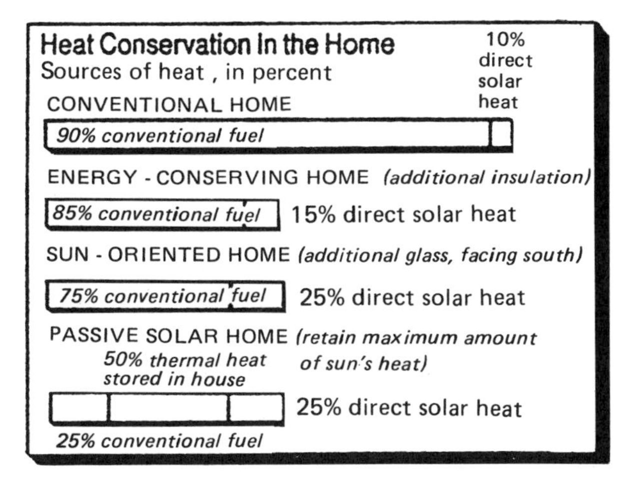

Figure 6.17 How are we to interpret this uniquely confusing plot about heat conservation? (From Reference 11, copyright 1983 by the New York Times Company. Reprinted by permission.)

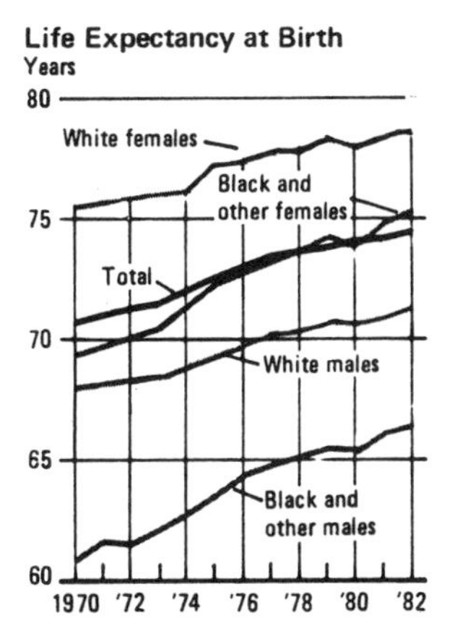

Figure 6.18 Small and confusing. How do we separate "Total" and "Black and Other Females"? (From Reference 12.)

bold line (‘‘Total’’) cutting across one of the others. This is a small graph and the type is small. It needs short labels which are easy to read. Why do we have to puzzle out the meaning of ‘‘Black and other females’’ to compare with ‘‘White females’’? How much easier it would be for the interested reader if the graph maker had identified the two lines as ‘‘White females’’ and ‘‘non-White females’’ so they could easily be identified as applying to complementary groups.

3. The Ultimate Graph

Some days Dr. Nowall feels completely frustrated. No matter how much he computes and plots he can't get a headline-worthy graph. What to do? If he is to attract the attention he wants, he needs some way to show a ‘‘soar’’ or a ‘‘dive’’—something which looks exciting or threatening to people who are reading on the run. On one of those black days, he thought up the following procedure reported by Black [13].

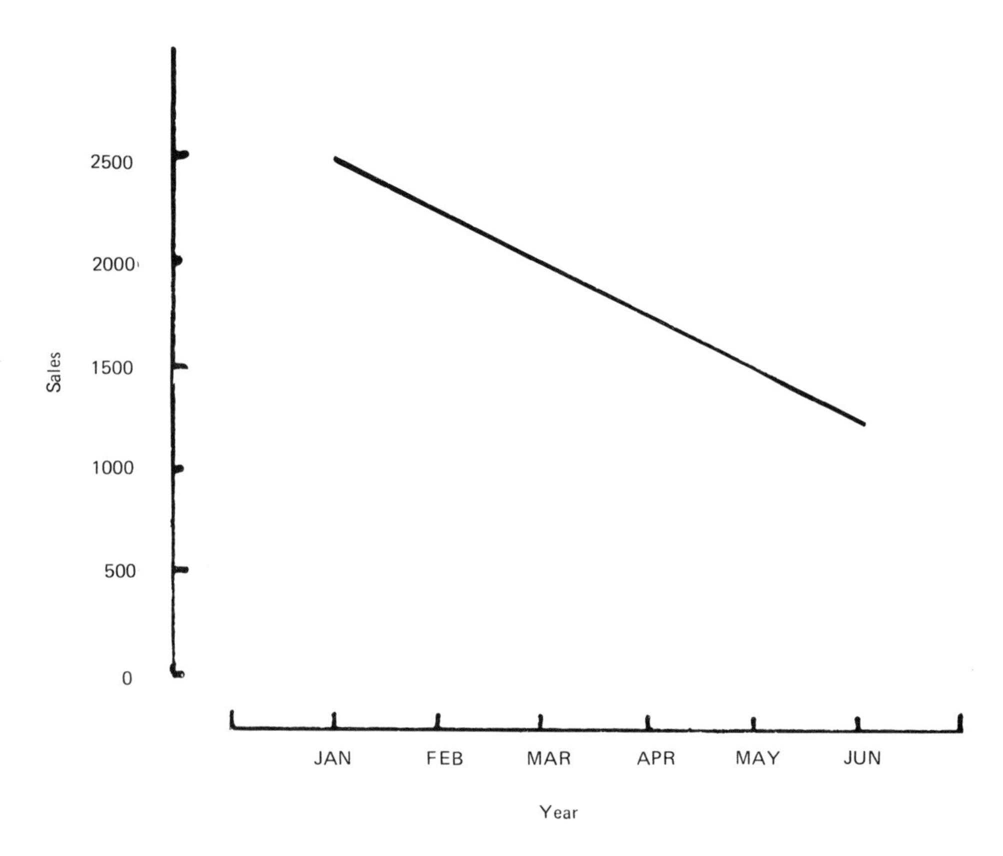

Figure 6.19 Depressing monthly sales.

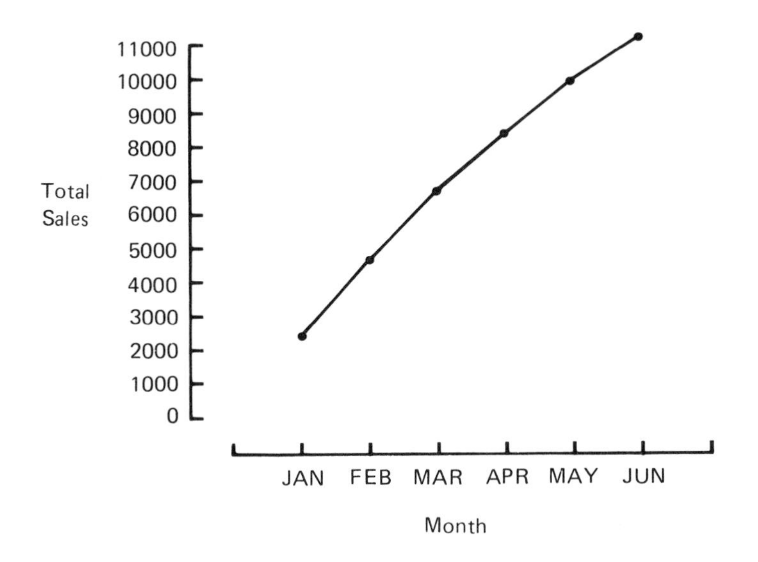

Figure 6.20 Encouraging cumulative sales from the monthly sales of Figure 6.19.

Suppose that a company experiences declining monthly sales starting from a high of 2500 units in January and declining to 1250 units by June.* The decrease is substantial, 250 units per month, as shown in Figure 6.19. Dr. Nowall consults to this company and is embarrassed by this slump; he sees his continued employment in jeopardy. To keep up management's morale, he plots *cumulative* sales as shown in Figure 6.20. Thus, good cheer can be taken from the report that in January 2500 units were sold and that 11,250 units have been sold by June. Pity the poor viewers of this graph who did not notice the difference between "sales" and "total sales."

III. Summary: Neither Sinned Against Nor a Sinner Be

The rules are so simple! You can easily avoid being a graphical sinner. Similarly, you can easily spot misuses if you know the cardinal sins:

Missing zero. The vertical scale should go to zero, or show clearly that it does not, by a break in the axis. If you want to show changes in some data, then plot the changes.

Distorted horizontal axis. The horizontal axis should be complete, with no breaks and no midstream changes in scale.

---

*These are contrived data based on a real occurrence.

Missing baseline. If the data are values of an index, show the baseline.

Distortion by use of an arithmetic vertical scale for economic time series. Any series of data which is growing at a constant percentage rate per year will "soar" when plotted against an arithmetic vertical scale. Replot any such data on a ratio scale. If you are plotting data for an economic time series, or you have other reason to be interested in percentage changes, then use a ratio scale on the vertical axis.

Inflated dollars. If the data you are plotting are monetary values plotted in current (actual) values, then replot them in deflated (or constant) values to determine their value in purchasing power and label appropriately.

False proportion. If the purpose of a picture of an object is to show a relative amount, check that the illustrated relative proportion is valid.

Visual confusion. Don't make confused or confusing graphs, and refuse to accept them in other's work.

Confusing labels. Don't make confused or confusing labels, and refuse to accept them from others. Apply the standards of clear communication and correctness to the labels of a graph.

Crowded to the point of incomprehensibility. Don't do it, and don't accept it.

Great differences in the lengths of the axes. Again, don't do it, and don't accept it. There are many informed opinions and an immense literature on the perception of form. Fred H. Knubel, director of Public Information for Columbia University, basing his opinion on many years of experience presenting information in graphical form, has concluded that the least biased presentation is an approximation to a square. Edward R. Tufte concludes his book on graphics with this statement, based on his study of the literature: "If the nature of the data suggests the shape of the graphic, follow that suggestion. Otherwise, move toward horizontal graphics about 50 percent wider than tall" [2, p. 190].

Remember the authors' definition of a misuse: Using numbers in such a manner that—either by intent, or through ignorance or carelessness—the conclusions are unjustified or incorrect. Scrutinize your graphs to reduce the likelihood of a reader drawing an unjustified or incorrect conclusion.

7
Methodology: A Brief Review

A workman is known by his tools.

Anonymous proverb

The king lost his way in a jungle and was required to spend the night in a tree. The next day he told some fellow traveler that the total number of leaves on the tree were "so many" (an actual number was stated). On being challenged as to whether he counted all the leaves he replied, "No, but I counted the leaves on a few branches of the tree and I know the science of die throwing."

From the ancient Indian epic Mahabharat
(Nala-Damayanti Akhyān)

I. Introduction

There are many useful tools—statistical methods—for analyzing data. At the most basic level, we use the pencil and paper tools of arithmetic. At the other end of the spectrum, we use procedures that may call for supercomputers.

Like all tools, the effectiveness of the tools of statistical analysis depends on using them as their designers intended. If you try to dig a post hole with a snow shovel, you only make a cavity which cannot hold a post and is dangerous to passersby.

The analogy with the snow shovel is appropriate. The designer of the snow shovel assumes that you are trying to make a shallow, wide swath (not a deep hole of small diameter) in a light, finely textured, loose material (not heavy, irregular, clumping soil). Similarly, the creative individuals who develop statistical methods to analyze data must make assumptions about both the user-analyst's purpose and the nature of the data, so that the tools they devise will be suitable to these constraints.

The analyst's purpose is to:

- Explore the data, looking for suggestions of future directions
- Summarize the data to make it comprehensible to self and others
- Evaluate the effect or effectiveness of some approach, method, treatment, compound, program, and so forth
- Compare two or more approaches, methods, and so forth

The assumed nature of the data may be that they:

- Are from a population having a particular distribution of values
- Have no measurement error
- Come from a situation in which the causes of variation are unchanging
- Are appropriately randomized
- Have a known variability
- Have the same variability as other data in a comparison, or do not change variability as some factor of interest changes
- Are from a random sample

The analyst's purpose and the nature of the data determine the statistical tool to be used.

II. Statistical Tools

The number of statistical tools is great and increases daily. However, only a few of them are used in the majority of reports and studies which reaches the public. In this section, we review several of the most common tools on which statisticians (and others working with statistics) rely.

Even basic arithmetic tasks such as counting, adding, subtracting, taking percentages, ranking in order, making plots, and so forth, can be basic tools to help us to understand the messages hidden in the data and communicate them to others. More complex tools are often concerned with summarizing data so that we can draw some conclusions without looking at the data in detail. Sometimes we learn more from the summaries than we can by working with the actual data. Examples of such tools are mean, median, standard deviation (a measure of the scatter, or dispersion, of the data), correlation coefficient, and coefficients in regression equations. In experiments, we are concerned with the design and analysis of the experiment by which we obtain the data; selecting the wrong, or inappropriate, tool is a misuse which can lead to invalid conclusions.

Incorrect or improper use of even a tool as simple as the mean can lead to serious misuses. For that reason, we start with a preliminary discussion of a few of the simple and most-used tools of statistics.

A. On the Average . . .

The most commonly-used statistical summary measure is a "typical value" for a set of data. Why would someone want a typical value for a set of data? The reasons are many. An engineer might want to have a typical value of the weight which causes a beam to deflect too far for safety. A series of tests are run on 100 sample beams and values between 100 kg and 150 kg are obtained. Should he report all 100 values? Should he report the middle value, 125 kg, to give the user an idea of what force is involved? But suppose most of the values fall near 150 kg?

Because they think of the values of data as falling along a line of values, many statisticians call a typical value a measure of "location," which tells where, along some imaginary axis, the values fall. In the case of the beam, values could be anywhere along an imaginary axis from zero to infinity. But what is the "location" of the values reported? Somewhere between 100 and 150; hence the concept of location.

Other statisticians prefer to think of a typical value as a measure of "central tendency," showing where the data tend to be, or cluster. For example, if the data were mostly near 125, and roughly symmetric above and below 125, the central tendency would be near 125.

For another example, consider the data in Table 7.1, listing the annual salaries of ten business executives. We will look at two of the statistical tools (which most statisticians would call measures) that can help us to get a typical value for these data.

Table 7.1 Annual Salaries of Ten Executives (in thousands of dollars)

Raw data
89
110
146
142
200
143
152
110
240
168

1. The Mean

The *arithmetic mean*, usually just called "the mean" or "the average,"* is the sum of all data values divided by the number of such values. Thus, for our executive salaries in Table 7.1, you compute the arithmetic mean by adding up all the salaries and dividing by the number of executives. In this case, the total for all the salaries is $1,500,000; divided by 10 you get a mean executive salary of $150,000. This is the value of salary that, if all executives in the study made the same salary, would give the same total value.

The arithmetic mean has the most meaning when the values are closely centered. But suppose that the one executive who earned $146,000 earned $646,000 instead. While most of the executive salaries are still "around" $150,000 and only one other makes more than $200,000, the mean has jumped from $150,000 to $200,000, an increase in the value of the mean of more than 30%. Clearly, the mean is not a good indicator of "typical values" in this type of situation.

2. The Median

The next most common measure of central tendency is the *median*. To find the median, take all the numbers you have collected and order them by increasing value. Once the numbers have been so ordered, the median is the middle value (if the number of values is odd) or the average of the two middle values (if the number of values is even). To get the median of the salaries in Table 7.1, order the values as shown in Table 7.2. Then find the middle value (or as in this case, the average of the middle two values) to get a median executive salary of $144,500 ($143,000 + $146,000 divided by 2). Note that the median is only a little less than the arithmetic mean.

Thus, the median is not much affected by widely dispersed values. For this reason it is often used for reporting typical salary, age, and similar values. The median is also useful because it is easily understood—it is that value such that about half the population has values below, and half has values above, the median value. In 1983, the median age of the U.S. population was about 31 years; about half of the U.S. population of 234 million people were younger than 31 years, and about half were older [2, Table 5].

As we show above, the mean is greatly influenced by extreme values. On the other hand, the mean can be appropriate for many situations where extreme values are not obtained. To avoid misuse, it is essential to know which summary measure of the data to use and to use it carefully. Understanding the situation is necessary for making the right choice. Know the subject!

---

*Technically, this is a misnomer. For example, the U.S. Census Bureau says, "An *average* is a number or value that is used to represent the 'typical value' of a group of numbers" [1, p. xvii].

Table 7.2 Annual Salaries of Ten Executives, Arranged in order of Increasing Magnitude (in thousands of dollars)

89
110
110
142
143
146
152
168
200
240

B. Rate and Counter-Rate

The *rate* is another simple, frequently-used statistical tool. You apply it when you wish to know: (1) the relative magnitude of some quantity; or, (2) track its change with time. The U.S. Bureau of Labor Statistics determines the unemployment rate by measuring relative magnitude; this makes newspaper headlines every month. It is an example of the first kind of rate, where a value, the number of the unemployed, is divided by the size of the whole group, the labor force.* For example, if 10 million employable people are reported as unemployed in a given area at a given time and there are 100 million people in the labor force of that area, the unemployment rate is 10%.

If the unemployment rate rises to 11% in the next month, newspaper headlines say "Rise in Unemployment!" But every rate of this kind has a "counter-rate," a term rarely seen in textbooks or media. If the unemployment rate is 10%, then the employment rate is 90%; if the unemployment rate rises to 11%, the employment rate falls to 89%. Few headline writers would find a change in the employment rate from 90% to 89% significant or newsworthy.

Does this seem to be a petty issue? This principle (rate and counter-rate) has recently been applied in quality control. Most quality control systems control the work of people and machine processes by reporting on the rate of defectives: how many defective items are produced out of a lot of a given size. Typical values of the proportion defective in practice run from as low as .000001 to .1 or .15. There is a morale factor here in that the workers are constantly being told

---

*See Section III.B of Chapter 8 ("Would a change in definition make any difference?") for an explanation of the definition of "labor force."

"how bad" their work is. Some quality control engineers achieve improved morale and better performance by concentrating on improving the counter-rate, the proportion "good" [3]. Thus, a group of workers may be trying to raise the quality of their work from 99% to 99.8% good, instead of decreasing the proportion defective from .01 to .002.

We cannot, unfortunately, provide a complete catalogue of statistical tools, but the median, rate, and counter-rate are tools which the nonprofessional can easily comprehend when evaluating a statistical analysis. Below we illustrate these and other tools using specific examples, some real, some hypothetical.

III. Some Preliminary Examples

A. Sometimes It's Hard to Count Right

As we mentioned in the introduction to this chapter, even simple arithmetic is a tool, a method of statistics. And it is sometimes hard to count right, as you shall see. True, our first example involves counting with a computer. In general, however, the computer just makes counting easier and more accurate. But not always.

In 1974, Dr. Martin S. Feldstein, who later served as chair of President Reagan's Council of Economic Advisors, produced the results of a massive data analysis by computer which showed that Social Security deductions reduced personal savings by 50% and made the country's plant and equipment 38% smaller than it would have been without Social Security [4–6]. In 1980, "Professor Feldstein acknowledged [a] mammoth mistake . . . [which] led to a multi-billion dollar overestimate of the negative effect of Social Security on national saving" [4].

Do economists often make mistakes of such magnitude? Alan Blinder, an economist, says

> There are probably untold numbers of errors buried in the economic literature. . . . If you had made a small programming mistake . . . it would probably not be discovered unless it had produced crazy numbers, and Marty's did not. [5]

In fact, "Marty's" calculations did produce results which, if not "crazy," were strange enough to arouse the suspicions of two other economists, Dean Leimer and Selig Lesnoy. They decided to check Feldstein's results, since they knew that: "Other major studies have yielded smaller estimated effects on saving or concluded that there was no evidence of a significant effect" [6].

Why wasn't Professor Feldstein more suspicious? We can't say for sure, but perhaps it was his strong belief in his conclusions, even if erroneously obtained, for after the "mammoth" error was found, he stated: "My sense is that there is a general belief in the profession that Social Security still depresses savings, although the evidence is not finally in on the magnitude of that effect'' [5]. He has also argued that: ''When [a] legislative change is taken into account (and the error in the Social Security wealth series is corrected) the results are very similar to the conclusions reported in my earlier study'' [8]. The disagreement among experts and the results of analysis so far lead to only one reasonable conclusion—that we do not know the effect of Social Security on saving. But the strongest and most widely-publicized conclusion on this issue was based on an analysis with a ''mammoth'' computational error.

B. Divorce Is Good for Your Health

A graphical tool, the *scatterplot* is a plot of two variables which can be assigned to a series of points plotted at positions corresponding to the value of one variable on the horizontal axis and the value of the other on the vertical axis. We describe its use in the fictitious example which follows. In this section, we also introduce the correlation coefficient which involves computations on the data. To make our point about these tools clear without the need for an extensive explanation of the subject matter, we use a fictitious example. Unfortunately, many real misuses of these tools are as excessive and bizarre as our little piece of fiction.

Dr. K. Nowall's success in demographic analysis led him to investigate divorce rates, a matter of considerable public concern and certainly worthy of a few television interviews if he could develop some important results. Browsing through the *U.S. Statistical Abstract*, he found some suggestions in the data on death rates and divorce rates that he decided to investigate in detail.

Death rates in the United States are different in various regions as defined by the U.S. Bureau of the Census and shown in the *U.S. Statistical Abstract* [9]. The divorce rate, the number of divorces and annulments per 1000 of population, also varies among regions. Dr. Nowall studied the data and saw a relationship between death rate (deaths per 1000 of population) and the divorce rate (number of divorces and annulments per 1000 of population) for 1980 [9, Tables 106 and 122]. He further investigated this belief by making a *scatterplot* of these data.

To make his scatterplot, Dr. Nowall measured the divorce rate along the horizontal axis and the death rate along the vertical axis as shown in Figure 7.1. Each point in Figure 7.1 corresponds to a given region of the United States as defined by the U.S. Bureau of the Census; its position on the scatterplot is determined by the values of the two variables for that region.

Dr. Nowall examined this plot and saw a positive relationship between the death rate and the divorce rate. By ''positive,'' he meant that as the divorce rate increased, the death rate tended to decrease. He used the term ''tended'' because the points did not lie on a straight line of direct proportionality.

The strength of an association is measured by the *correlation coefficient*,

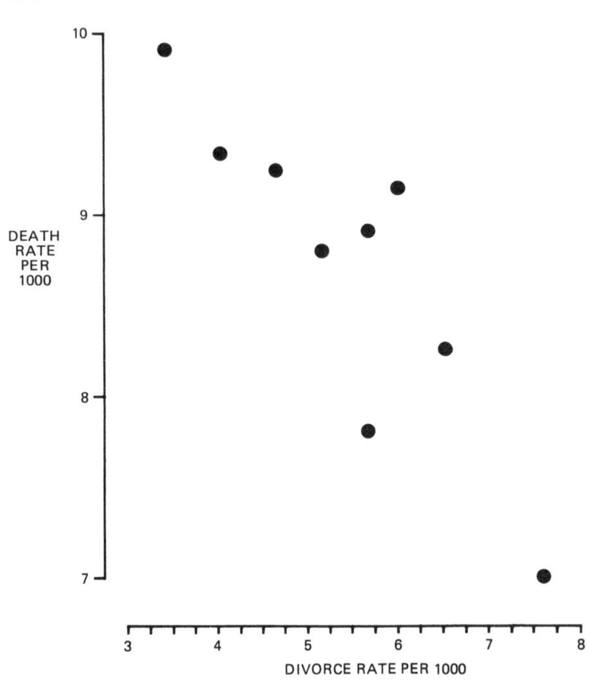

Figure 7.1 Dr. K. Nowall's scatterplot of death rate and divorce rate for the regions of the United States.

another analytical tool requiring computations which are tedious if done with pencil and paper, but easy to get with a computer or a calculator. The correlation coefficient ranges from -1 through 0 to $+1$. The correlation coefficient may be calculated by a number of different methods; each may give a different value for the same data. The "Pearson product-moment correlation coefficient" is most commonly used and is denoted by "R" or "r." This is the correlation coefficient which Dr. Nowall computes using his pocket calculator.

A value of $R = +1$ is an association such that points lie on an upward-going line (as shown in Figure 7.2), and $R = -1$ to a similarly tight association on a downward-going line (as shown in Figure 7.3). Examples of intermediate values of R are shown in Figures 7.4 and 7.5. Figure 7.5 shows a "loose" downward-going association between two variables. Dr. Nowall gets a value of $R = .869$ from his data, which he feels is quite good. After all, he has seen many research papers which draw important and influential conclusions from data having much lower values of R.

Dr. K. Nowall's scatterplot and value of R show that the values of death rate and divorce rate for regions are associated. That is, they seem to be paired

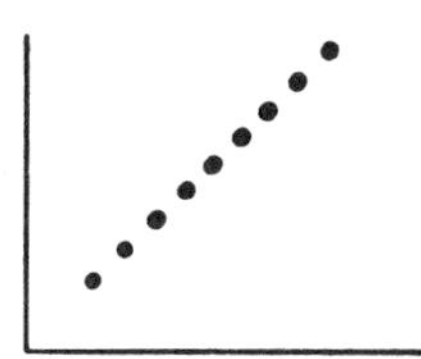

Figure 7.2 Scatterplot showing perfect positive correlation ($R = 1.0$).

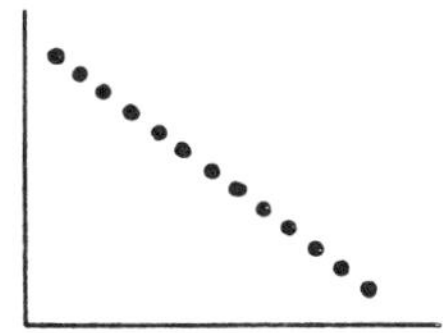

Figure 7.3 Scatterplot showing perfect negative correlation ($R = -1.0$).

Figure 7.4 Scatterplot showing less than perfect positive correlation (R is between zero and one).

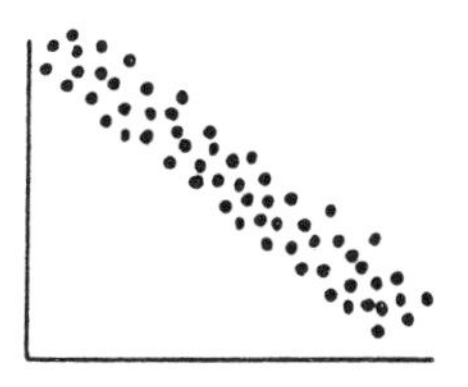

Figure 7.5 Scatterplot showing less than perfect negative correlation (R is between zero and minus one).

together in a decreasing (negative) relationship. Association tells us that much but affirms no more. Unfortunately, Dr. Nowall is not too perceptive and when he looks at Figure 7.1, he is sure he knows what causes high death rates in particular regions.

Dr. Nowall believes that Figure 7.1 supports his belief that divorce is good for your health: the greater the number of divorces per 1000 persons, the lower the death rate. He leaps from association to *causation,* and concludes that divorces *cause* a reduction in deaths. In his press release, he suggests a systematic policy of increasing the numbers of divorces in regions with high death rates, by offering financial incentives, training married couples in divorce law, and providing divorce instruction to all couples receiving marriage licenses. He also promises to start a study of the effect of divorce on the number of inmates in mental hospitals.

Like many others, Dr. Nowall has misused statistics by *confusing association with causality.* This is a common source of false generalizations about the behavior of people, organizations, countries, ethnic groups, and many phenomena in the physical world. To show just how badly he has misused statistics, we must give a short discussion—far from complete—on cause. John Tukey, one of the great statisticians of our time, says that to support the notion of "cause"—which Dr. Nowall tried to do in concluding that physicians cause death—we need *consistency*, *responsiveness*, and a *mechanism* [10, p. 261].

Let us imagine the following situation: The community of Hypertown finds that its drinking water contains a certain amount of a pollutant. Also, its inhabitants have, on the average, "too high" blood pressure. The health commissioner of Hypertown looks at surrounding communities and finds another community, Hypotown, which has none of this pollutant in its water supply. The average blood pressure of its inhabitants is lower than that of Hypertown residents. If this pattern is found in a number of other pairs of communities, then we have *consistency* of findings.

To confirm *responsiveness*, you must show that changing the level of pollutant in the water supply changes the average blood pressure, at least up to a point. There is no necessary upper limit to how much pollutant there may be in the drinking water unless the "water" consists solely of pollutants, in which case it is no longer water. But there is certainly an upper limit to blood pressure for specific individuals. If it gets too high, the person dies.

In general, to confirm responsiveness you must run an experiment. In this case, you would like to have an experiment in which the amount of the pollutant fed to human subjects was varied and their blood pressure monitored to confirm or disprove that the rise and fall of blood pressure is directly related to the amount of the pollutant consumed.

To support the existence of a *mechanism*, you must be able to construct a model of the real world based on known principles which can, perhaps step-by-step, explain the effect. In the example we are considering such a mechanism could involve showing that the pollutant directly acts on cells in the kidneys to affect the retention of sodium and water in ways that have been shown through medical research to produce an elevated blood pressure.

To prove a *causal link* you must show that the presumed effect does not cause the presumed cause. In our example of the pollutant and high blood pressure, if we did not have a clearly demonstrable mechanism, we would also have to show that it was not something about people with high blood pressure that makes them drink water containing the pollutant.

Thus, if you have no more than a statistical demonstration of association between two variables (such as death and physician rates) you have no basis for saying that one causes the other. You do have a basis for suspecting that this might be so and then working to confirm or disprove your suspicion.

As for Dr. Nowall, he must show—through consistency, responsiveness, and a mechanism—that divorce reduces the likelihood of death. Do you think he will succeed in doing so?

C. Will Married Couples Sacrifice Their Marriages to Reduce the Death Rate?

Linear regression is a complex statistical tool.* Even for data sets of modest size, the computations are tedious to perform with pencil and paper. But we can now carry out these computations with computers and calculators. The tedium has been eliminated but difficulties in interpreting the results remain, as we show by continuing our fictitious, but not unrealistic, example.

Alas, Dr. Nowall is not familiar with the concepts of causation and continues to pursue his idea of reducing the death rate by selectively increasing the divorce rate. By how much does he have to increase the number of divorces in the New England region to equal the (lower) death rate in the Mountain region? Or, given that the increase in divorces must be accomplished over a period of time and at some cost, what reduction in death rate can be expected for a given increase in divorce rate? To what level must the divorce rate be raised to push the death rate to zero?

If he could draw some "best" straight line through this set of data, he could answer all of these questions by using the straight line to extrapolate values of death rate for given divorce rates. He can easily do a linear regression on his pocket calculator. The method used on most pocket calculators and computer programs to do linear regression gets the formula for the straight line through the

---

*We do not have enough space to fully explain linear regression to readers who are unfamiliar with this tool. Such readers can "skim" these sections and still grasp the essence of the misuse we discuss.

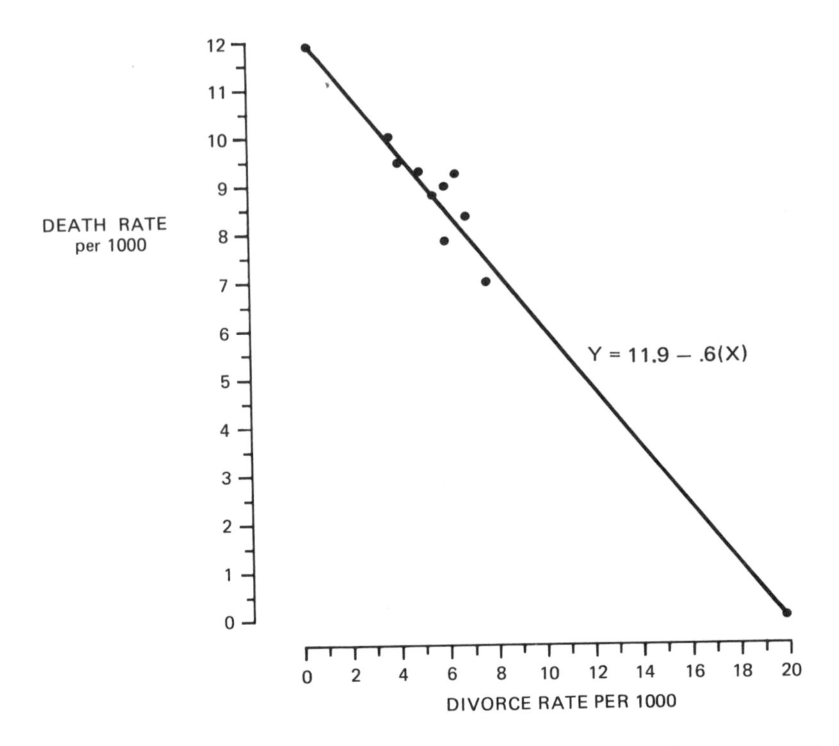

Figure 7.6 Dr. K. Nowall's scatterplot of death rate and divorce rate with regression line for the data.

data that gives the least value for the sum of the squares of the "residuals" (vertical distances between the line drawn and the plotted points).*

Nowall enters the data, the calculator does the analysis and reports that the regression line through the data is given by $Y = 11.9 - .6*X$ which is shown in Figure 7.6. The scatterplot of Figure 7.1 has been redrawn to show $X = 0$ so that you can see that the regression line has the "Y-intercept" value of 11.9; this is the value of the death rate for $X = 0$, a divorce rate of zero, which means that there are no divorces in the region. He now finds the value of X for $Y = 0$. This is $X = 11.9/.6$ or about 20 divorces per 1000 of population, a small price to pay for no deaths in the region. For those who are interested in less dramatic improvements, he announces that since the multiplier of the divorce rate is .6, for each 1.7 divorces (1/.6) one person will not die.

Has Dr. Nowall proved that divorces cause deaths, or only that, for these

*And is, for that reason, sometimes known as the "least squares line."

regions high values of each variable are associated. To prove cause and effect, he must show consistency, cause, and mechanism. Can he prove this, when there are several other variables which may be causal factors in the relationship between divorce and death rates? One obvious variable is the age distribution of regional population. We have good reason to suspect that older married couples are less likely to divorce and that they are more likely to have a higher death rate than younger married couples. The jury is still out, and may stay out for a long, long time.

Now let's look at a real-world use of this methodology. According to a report in the July 1986 *Scientific American* (p. 62), Gregory B. Markus of the University of Michigan has found that "The concern of political parties for the 'image' of their presidential candidate is largely misplaced. . . . Voters," he said, ". . . make up their minds primarily on the basis of personal and national economic conditions." He came to this conclusion by examining national economic statistics in conjunction with data on national elections from 1956 through 1984.

"Markus estimates that each increase of 1 percent in disposable income increases the vote for an incumbent by 2.2 percentage points, other things being equal," according to the article. He points out that 1980 was the only presidential election year since 1952 when the real disposable income declined. In that presidential election, incumbent President Jimmy Carter lost to Ronald Reagan. "Markus estimates that each increase of 1 percent in the disposable income increases the vote for an incumbent by 2.2 percentage points, other things being equal." Markus is quoted in the article as saying that a close analysis of that election "provides no evidence for the contention that Reagan's victory was the result of his policy or ideological positions." Do you believe it?

IV. More Examples

A. Predicting the Unpredictable?

People want to know what the future holds. *Extrapolation* is a statistical tool which has been of considerable value, particularly in science and engineering. But it can also be treacherous, especially in the hands of the reckless, the unskilled, and the sensation-seeking. We continue with another fictitious example.

Dr. Nowall knows that a spicy statement predicting the future will attract attention, perhaps some headlines, and maybe even an exciting five minutes on a radio or TV show. He believes that he can use the same general procedures which he used to relate death rates to the prevalence of divorce to "predict" the future of other important demographic entities.

Many methods, such as linear regression, are used to project future values. In certain circumstances (control of inventories, early warning of cost overruns, and so forth) they can be useful, but all have problems of accuracy because no method can reveal the "truth" about the future. Careful prognosticators refer to "projections" and not "predictions." Only headline seekers label their guesses as "predictions." Dr. Nowall is oblivious to these cautions.

1. *Trot Out the Prams*

School boards planning school construction projects, economic planners concerned with public policy, and certain business strategists wishing to direct market actions based on future births and population size are interested in predicting birthrates. In 1959, Dr. Nowall saw this as fertile ground for statistical analysis. He observed that the birthrate per 1000 of U.S. population slowly climbed (with some intermittent decreases) from 20.4 per 1000 in 1945 to 25.3 per 1000 in 1957. Dr. Nowall saw that the birthrate dipped a bit in 1958 and 1959, but to his experienced eye this dip seemed insignificant. After all, there was a similar dip from 1948 to 1950, after which the post-World War II upward trend resumed.

In 1959, his goal was to project what the birthrate would be in 1969 so others would use his "predictions" in making demographic forecasts. Using pencil and paper, he quickly computed from the 15 pairs of data for the years 1945 to 1959, the linear regression line for birthrate (Y) as a function of the two-digit year (X) to be $Y = 19.71 + .094X$.* Setting the value of X to 69 (for 1969) in this formula, he projected a value of the birthrate for 1969 of 26.2 per 1000. Alas, the actual birthrate in 1969 turned out to be 17.8 per 1000! The dip in 1958 and 1959 was the beginning of a downward movement.

Figure 7.7 shows Dr. K. Nowall's work in this instance. Using the 15 sets of actual data from 1945 to 1959, he computed the regression line shown, which tips upward slightly. As you can see from the plotted points, there really wasn't much of an upward trend during that period, but the low post-World War II value for 1945 "pulled" the computed regression line downward at the left end. It is an artifact of simple least-squares linear regression (as is almost invariably used in these cases: the function is available on pocket calculators and in computer programs) that the line is strongly influenced by values which are extreme compared to the majority of values. Thus, when he extended the straight line ahead ten years to 1969, he computed a birthrate greater than all but one of the 15 birthrates used to obtain the regression line.

Should he have recognized that the values for 1958 and 1959 were the beginning of a downward movement? If he knew his subject (see Chapter 3) he would have known that birthrates respond to social and economic factors. The low birthrate of 1945 is to be expected with so many men at war or just coming home, combined with the uncertainty of wartime. The increase in birthrates in 1946 and 1947 is to be expected as military personnel return home, form families

*The raw data come from Reference 2, Table 80.

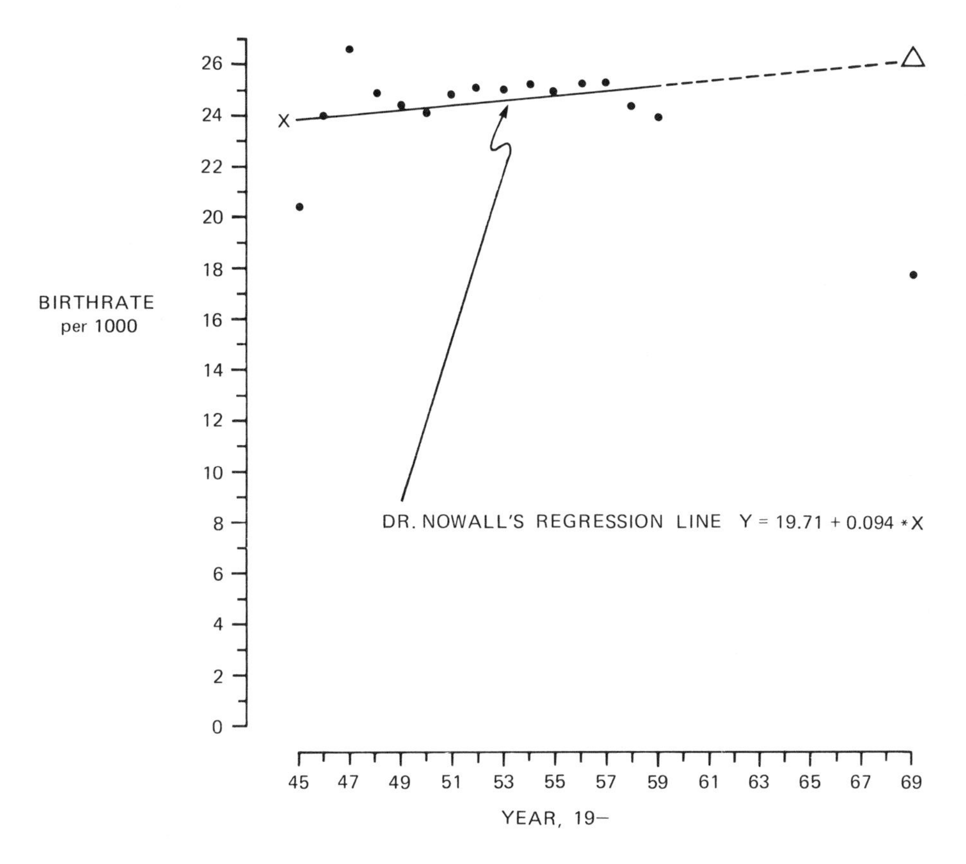

Figure 7.7 Actual birthrates for 1945 to 1959 and for 1969, and Dr. K. Nowall's linear regression line and projection of the birthrate for 1969.

and plan their future lives. In a preliminary analysis, the period 1949 to 1957 can be regarded as a period of relative stability in the process of family formation and growth. In 1959 it would have been premature to regard the two successively lower values in 1958 and 1959 as the beginning of a downward movement (as we can now see it was, in hindsight). But those two values should have been enough to warn Dr. Nowall of the danger of extrapolating the straight line whose upward trend was largely due to the special situation of 1945.

2. Who Will Be a Good Student?

We administer tests to millions of students every year in order to make decisions as to whether to admit them to colleges and universities. In general, the purpose of admissions policies is to accept candidates with a high probability of success at the admitting institution. The purpose of the scores on standardized

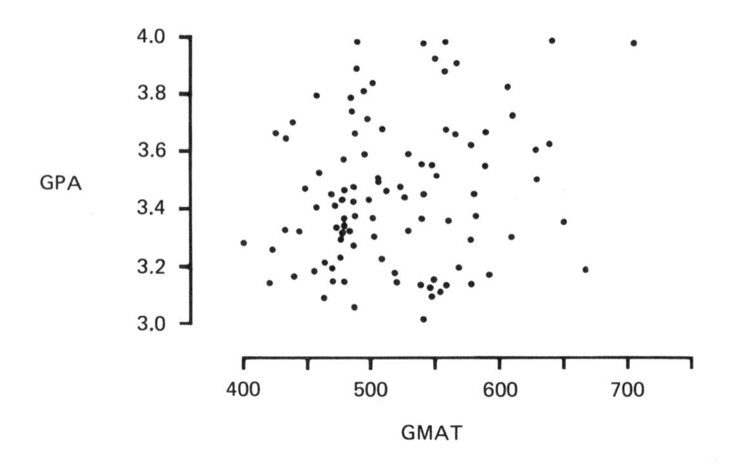

Figure 7.8 Scatterplot of MBA Grade Point Average (GPA) and Graduate Management Admission Test (GMAT) score for 100 students in an MBA program.

tests such as the Scholastic Achievement Test (SAT), Graduate Management Admissions Test (GMAT), Graduate Records Examination (GRE), and so forth, is to help admissions committees predict probability of success, one of the major inputs in making the admission decision.

In statistical terms, the goal is to use the test score to predict the Grade Point Average (GPA) the student will earn in the admitting institution. Any institution (or group of institutions) can get data on test scores, and the GPA statistics are from the records of currently enrolled students.

One set of such data is shown in Figure 7.8, which shows the scatterplot of GPA versus GMAT* for 100 students enrolled in a particular Master of Business Administration (MBA) program. The correlation coefficient for these data is $R = .21$, which drops dramatically to $R = .16$ if the student in the upper-right corner is eliminated from the computation. This one exceptional student has a large effect on the perceived association. This institution uses the GMAT as the major component of its decision on whether to admit a student! Would you call this a misuse of statistics?

Before making a judgment, there is one important fact to consider: Since the GMAT was used as a criterion for admission, no applicant with a GMAT score below 400 was admitted. Thus, we have no way of knowing what GPAs individuals scoring below 400 on the GMAT would have gotten if they were students. It is possible that if we had these data we would find that there was a much higher correlation between the GPA after admission and the GMAT score

*One of the criteria for admission used by many business schools.

before admission. Four possible distributions (including the students below the cutoff value) are shown in Figures 7.9–7.12. Which do you think is most likely? Our use of these tests is based on the belief that the downward trending distributions of Figures 7.10 and 7.11 are the likely distributions.

It is a misuse of statistics to conclude that a student's GMAT score is a good predictor of the student's GPA *based on only the data of Figure 7.8*; you can draw that conclusion only if you know that the whole distribution (including students with scores below 400) is similar to Figure 7.10 or 7.11. Nor can you conclude that it is a *bad* predictor in the absence of knowledge that the whole distribution is similar to Figures 7.9 and 7.12. This is a statistical "Catch-22." Schools admit students based on the assumption that the test is a good predictor of future academic performance. They do not admit students below a certain test

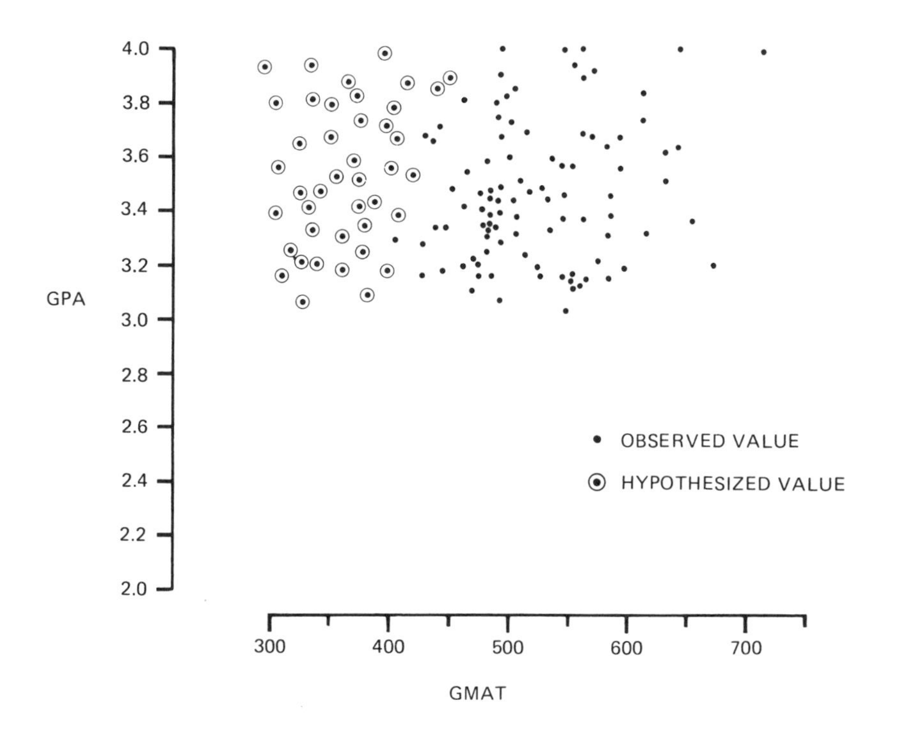

Figure 7.9 Hypothesized scatterplots of student performance versus admission test score are shown in this and the following three figures. The points above a GMAT score of 400 are observations from Figure 7.8. The GPA values corresponding to GMAT scores below 400 are hypothetical values that might have been earned by those students who were not admitted because their GMAT scores were below 400. Which of these scatterplots do you feel is most likely?

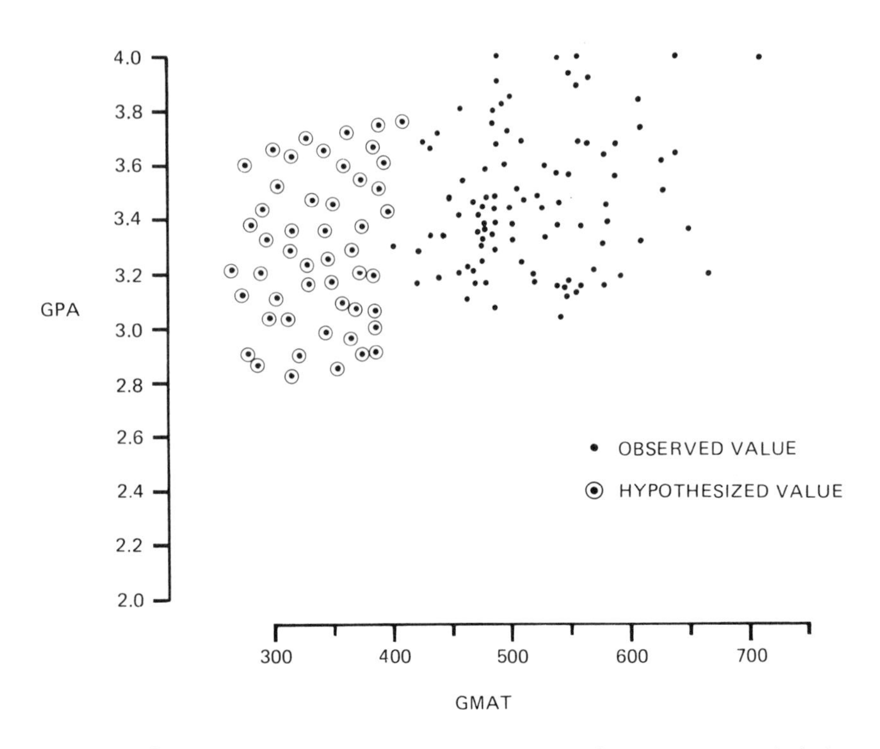

Figure 7.10 An hypothesized scatterplot of student performance versus admission test score.

score, and therefore do not get statistical data which would enable them to ascertain whether their assumption is correct.

The Educational Testing Service (ETS), which designs and administers the GMAT, summarized the results of studies at 20 business schools and found that the median correlation coefficient between GMAT scores and first year GPA was .35; and values as low as −.12 and as high as .76 were observed [11]. Unfortunately, we do not know the cutoff score for admission for the individual schools. However, the median value of .35 is some support for a belief that *in general* (as opposed to a particular case, such as the school illustrated in Figure 7.8) a higher test score predicts a higher mean GPA.

One way for a particular school to estimate the nature of the distribution below the cutoff score is to admit a small sample of randomly chosen students whose scores are below the cutoff score and track their performance. Of course, the same method could be used for a sample of schools to evaluate the validity of the test more generally. Another approach would be to compare the observed correlations between schools which have high cutoff values for test scores

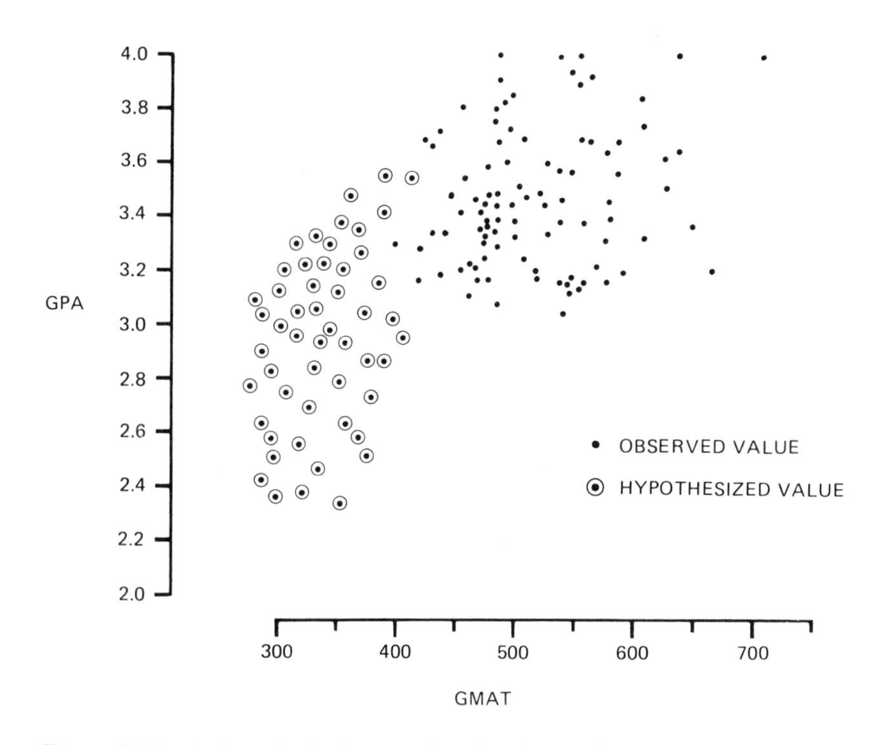

Figure 7.11 An hypothesized scatterplot of student performance versus admission test score.

(which, because of the cutoff of so much of the lower end, have low correlations) and schools which have low cutoff values (which will have a wider range of test scores and would show higher correlations if Figures 7.10 and 7.11 are the underlying distributions). We know from personal discussion and our review of publications that many admissions officials are aware of these issues and are working to resolve them.

3. *Giving Meaning to the Correlation Coefficient*

We now have enough background to return to our consideration of the correlation coefficient and give it some meaning. To some people, a correlation coefficient of $R = .2$ (the particular instance of the school illustrated in Figure 7.8) or $R = .35$ (the median value for 20 schools given in the ETS report) may sound significant in predicting one variable (such as GPA) from another (such as GMAT score). Fortunately, we can give tangible meaning to the correlation coefficient.

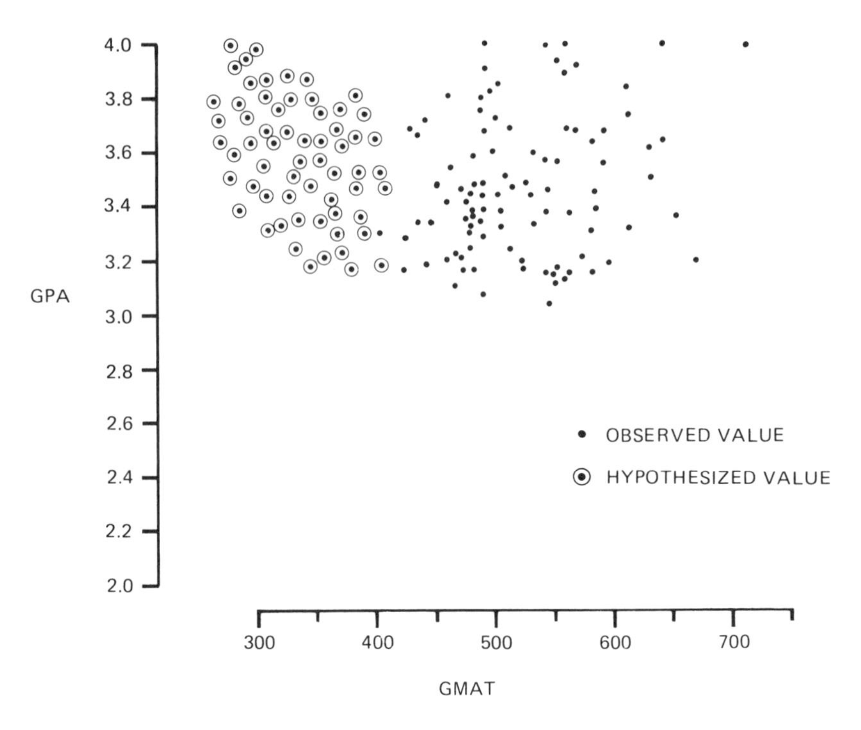

Figure 7.12 An hypothesized scatterplot of student performance versus admission test score.

Let's use a simple fictional business example to show how you can give meaning to the correlation coefficient. The Farr Blungit Corporation sells blungits through ten retail stores. The sales of blungits, in thousands, during the past month for each of the ten stores are shown in Table 7.3.

Farr plans to sell blungits through additional stores. *Using only these data* for the ten stores, what is a reasonable estimate for the monthly sales of blungits in a new store? In the absence of any other information about the store (such as location, the types of potential customers, advertising and sales promotion activities, seasonality, and so forth), you can start by using the average (mean) of these data—7600 units—for your first approximation. But how good is this estimate? The difference between this average and the observed values is quite large, ranging from −6800 to +8400 units!

The measure we have introduced for "spread" is the standard deviation, which in this case is about 5100 units. Statisticians and researchers using statistics often measure the dispersion with the square of the standard deviation,

Table 7.3 Last Year's Sales of Blungits at Ten Stores (in thousands of units)

3.2
7.9
1.8
12.6
12.1
0.8
5.8
10.6
16.0
4.9

called the *variance*.* In this case, the variance of the blungit sales data is the square of 5100 units, or about 26,000 units. To predict the sales of blungits in a new store so that you can plan manufacturing, distribution, and staffing of the store, you want to make the best estimate you can from the data. To put the concept of "best" into numerical terms, a statistician might say that Farr should reduce the spread, or variance, in this estimate from 26,000 to a lower value.

How might we reduce this variance? Is it possible that some variable such as those mentioned earlier (type of potential customers, advertising and sales promotion, and so forth) would explain away some of the variance and give us an estimate of annual sales of blungits with a smaller spread (hence, a lower variance)? Such a variable, whether it does or does not reduce the variance, is called an *explanatory variable*. Farr's marketing manager held a meeting to discuss this issue and the initial consensus was that advertising and sales promotion expenditures (ASP) were most likely to account for the high variation in individual store sales of blungits. Data on the magnitude of the explanatory variable, monthly advertising, and sales promotion expenditures were collected for the ten stores and are shown in Table 7.4 alongside the number of units sold.

These data are plotted in Figure 7.13 with the regression line that Farr's market researcher fitted to the data. You can look upon the regression line for sales in thousands ($Y = 1.8 + 4.8X$) as defining a *moving arithmetic mean* that depends on the ASP expenditures. Thus, for any level of ASP expenditure, you have some "mean" to use as an estimate. This mean depends on the value of ASP expenditures.† To measure how well the line estimates the observed values,

---

*Both the standard deviation and its square, the variance, are tools which summarize the amount of dispersion, or spread, of data.

†And, hence, is technically called the "conditional mean," since it is "conditional" on the value of the explanatory variable.

Table 7.4 Last Year's Sales of Blungits and Advertising and Sales Promotion Expenditures (ASP) at Ten Stores

Sales (in thousands of units)	ASP (in thousands of dollars)
3.2	0.0
7.9	1.0
1.8	0.0
12.6	2.0
12.1	2.0
0.8	0.0
5.8	1.0
10.6	2.0
16.0	3.0
4.9	1.0

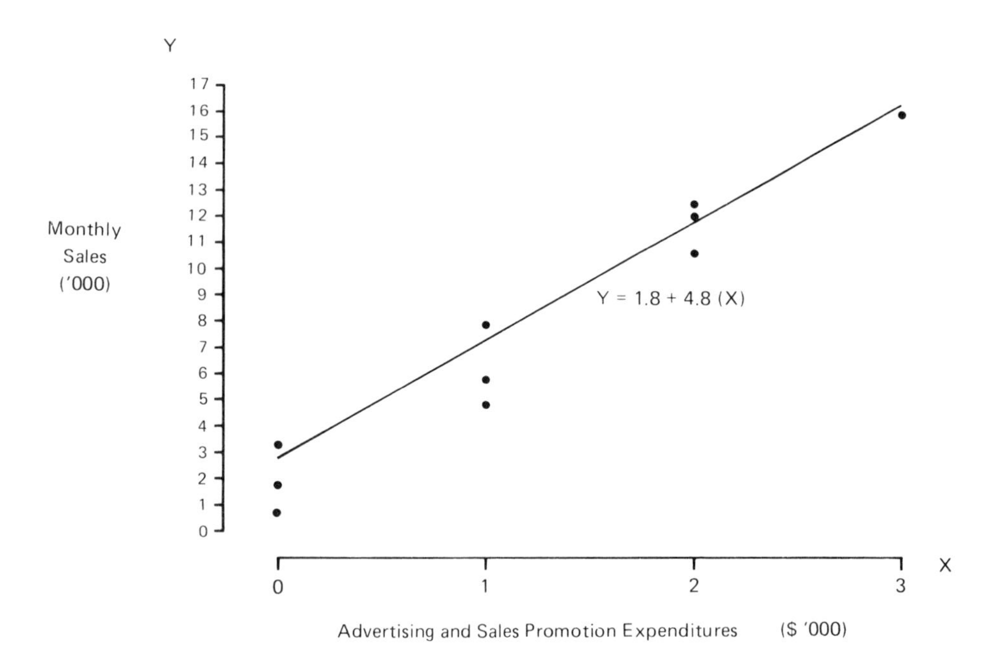

Figure 7.13 The scatterplot of blungit sales and advertising and sales promotion expenditures with its regression line.

Table 7.5 Residuals for Last Year's Sales of Blungits for Advertising and Sales Promotion Expenditures (ASP) at Ten Stores as the Explanatory Variable

ASP (in thousands of dollars)	Sales (in thousands of units)	Moving mean	Residual
0.0	3.2	1.8	1.4
1.0	7.9	6.6	1.3
0.0	1.8	1.8	0.0
2.0	12.6	11.4	1.2
2.0	12.1	11.4	0.7
0.0	0.8	1.8	−1.0
1.0	5.8	6.6	−0.8
2.0	10.6	11.4	−0.8
3.0	16.0	16.3	−0.3
1.0	4.9	6.6	−1.7

Note: Calculations from the exact equation $Y = 1.775 + 4.829\,X$ have been rounded to one decimal place.

we measure the variance with respect to this moving mean. For each observed value of sales, we can calculate the estimate given by the line. This is done in Table 7.5. The difference between the observed value and the predicted mean is called the *residual*. As you can see in Figure 7.13, the residuals around the regression line of sales are much smaller than were the residuals around the average sales. For any new store, we can decide how much we will spend on advertising and sales promotion and make a better estimate of sales than by just using the mean of all the values.

For example, if we decide to spend $1500 per month on ASP, we predict sales of $1.8 + 4.8(1.5) = 9000$ units per month, rather than 7600 (the mean of all the data). How good is this estimate? Well, you can measure the dispersion of an estimate obtained this way by comparing the variance around the moving mean with the variance around the mean of all the values without taking ASP expenditures into account. The variance around the moving mean is 1200, which you can compare to the variance around the mean of all the data of 26,000.* Thus, we can say that use of the explanatory variable, ASP, has explained $26{,}000 - 1200 = 24{,}800$ of the original variance. Statisticians and researchers usually express this in percentage terms; in this case they would say that "the use of the ASP expenditure as the explanatory variable has explained 24,800 divided by 26,000, or about 95% of the variance."

*The computation is not shown here.

What does all of this have to do with the correlation coefficient? A lot! For the square of the correlation coefficient, R-squared *is* the proportion of the variance that has been explained by the use of the explanatory variable.* Because of its importance and this usage, R-squared is called *the coefficient of determination*, which is another tool.

Now you have an easy-to-use tool for making your own judgment of the meaning of a particular correlation coefficient: Its square is the proportion of variance explained by the explanatory variable. Thus, for the data of Figure 7.8, the GMAT test score explains only R-squared = .21 times .21, or .044—less than 5% of the variance in the GPA!† Ninety-five percent of the variance is *not* explained by the GMAT. The median proportion explained for 20 schools [11] is R-squared = .35 times .35, or about 12%. In this case, 88% of the variance is *not* explained by the GMAT.

4. *Where Do You Get Those Ideas?*

Two researchers carried out a study of attitudes towards abortion of over 1000 Catholic Mexican-American women who were living in Los Angeles [12]. They divided the women into two groups, depending on whether their upbringing was in Mexico or the United States, and performed a complex regression analysis. Whereas Dr. Nowall and our admissions committees used equations with only one "explanatory variable"—the divorce rate, SAT score, or GMAT score—the researchers used several, such as religiosity, husband's occupational prestige, ethnicity, husband's education, own education, age, family size. The last four explanatory variables can be accurately defined. The others are obtained from questionnaires designed to measure these ill-defined concepts. Despite the inclusion of these other variables, the general idea is the same here as in our earlier examples: to use the data to establish the extent of association of some explanatory variable or variables with a *dependent* variable (e.g., level of education as associated with attitude towards abortion) and to get an equation for that relationship.

To measure how well explanatory variables predict the dependent variable when there are several explanatory variables, we have yet another tool, the *coefficient of multiple determination*, which can be interpreted in the same way as the simple coefficient of determination we have already discussed.

In this case—attitude towards abortion—we have a number of explanatory variables. The values of R-squared for the two groups of women and their several

---

*This is an approximation which gets better as the number of data values gets larger. The minor difference between this value and the exact value is not relevant in most practical cases, and we ignore it here to make it easier for the reader to understand the discussion.

†In this same school, another study showed a correlation coefficient as high as .4. Thus for this school, the GMAT explains, at best, about 16% of the variance in GPA. The correlation changes with time, as does the composition of the student body, faculty, and course offerings.

explanatory variables were .15 and .13. To judge these values, recall that Dr. Nowall got R-squared = .869 for his analysis of divorce and death rates. In this case the explanation of 13 to 15% of the variance was obtained not by using one variable, but by using several variables. Any one variable by itself does not explain much. Have these researchers really found a basis for saying what factors enter into the attitudes of these women towards abortion when they can explain only some 13 to 15% of the differences in these attitudes, or is this another poor use of statistics?

B. Can Smoking Be Hazardous to Your Face-Lift?

Our question, "Can smoking be hazardous to your face-lift?", cannot be answered definitively by making a survey of the kind we will discuss in Chapters 9 and 10. Questions like this are best answered by running an *experiment* in which the investigator exercises control over the subject of interest and the conditions pertaining to the measurement. In any given experiment you use several statistical tools.

For example, suppose a paint company develops a new formula for exterior house paint which it believes will last longer than other paints. The paint is marketed for two years. Then the company takes a random sample of purchasers and asks them if they think the paint has lasted longer than other paints they have used. If the results are in favor of the new paint, what do we know? Only the consumers' *perceptions* of the performance of the paint. Important as this is for marketing purposes—and it can be a most important contribution to future profitability—we really know very little about the performance of the paint. Why?

This survey lacks the element of *control*. We don't know if the consumers used this paint differently than the others against which they are comparing it. We don't know if their comparisons are with respect to paints used in different locations (the comparison paint used in the Canadian Arctic, the new version in Northern California). We don't know the types of surfaces on which the paint is used or the amount and type of additives, the methods of application, the number of coats, the differences in weather conditions, and so forth. You should be able to add many more factors to this list which could influence the performance of the paint and are not controlled.

What you really want for evaluation is an experiment in which control is exercised over the relevant factors. Most people regard it as just a matter of common sense that you would: systematically apply the paint in the recommended way to many different but likely surfaces; expose the paint in different localities and different directions in those localities; and measure the performance with some objective measurement of thickness, adherence, flaking, and so forth. They also understand that there should be *controls*, using earlier or competitive versions of paints treated in the same way. You may have seen such controlled experiments for paint and other surface treatments where patches of the competing surface coverings are made on the same panel of a surface material so that all conditions of exposure and surface are identical.

And even where the measurement is by machine, most people would again regard it as a matter of common sense that the person doing the measurement would not know which paint was which, so as to avoid unintentional bias. When an experiment is run in this manner, we say that it is *blind* and as we discuss below, there are several kinds of blindness to consider.

Continuing the paint example, you might also be concerned about the location of the comparison patches on the test surfaces. Should our new paint always be in the upper right-hand corner? There might just be some effect which favored or did not favor the upper right-hand corner (such as wind direction or impact of rain). Thus we think in terms of either *randomizing* the placement of the patches, or systematically assigning patches to different locations.

All of these considerations of control, blinds, randomization, systematic allocation to test situations, and many others we don't have space for, are tools of the field of statistics known as *experimental design*. These are relatively new tools. Philosophical underpinnings were developed in the 18th century, but experimental design as we now know it is largely a 20th-century development, and still evolving. But we do have some basic tools (and those we discuss are paramount) which have enabled us to draw sound judgments from statistical data. Much of the recent progress in medicine (particularly new drug and epidemiological successes) is based on the application of these principles.

Unfortunately, we don't know the answer to the question of whether smoking is indeed dangerous to face-lifts. We should, because a diligent plastic surgeon collected a great deal of data from his large number of patients. But the surgeon did not use the tools of experimental design well. As a result, insufficient data were collected for anyone to answer the question [13, 14].

"Skin slough" is an undesirable after-effect of a face-lift in which skin sheds because of inadequate blood supply, delaying healing and leaving more pronounced scars. The surgeon suspected that patients' smoking habits might cause skin slough. What is called for is an experiment in which a comparison is made between patients with similar physiological characteristics, differing only in their designation as smoker or nonsmoker. There must also be a clear definition as to what is a smoker or a nonsmoker (how many cigarettes a day defines a smoker, how long must a nonsmoker have been a nonsmoker, and so forth). The operating surgeon should not know whether a patient is a smoker. This could require special precautions, since smokers carry telltale odors and their skin and teeth are discolored. The surgeon diagnosing skin slough should not know which patients are smokers and should not be the operating surgeon. To provide a properly controlled experiment, it might even be necessary to use several surgeons to determine whether skin slough has occurred.

But, alas, the wrong tool was used. The plastic surgeon took a survey, and an inadequate one at that. Of 1186 of the surgeon's patients, the operating surgeon determined that 121 had skin slough, of whom 73 said they smoked more than one pack of cigarettes a day, 18 said they did not smoke, and 30 were either nonrespondents or their responses were lost. The 73 were classified as smokers, and we can only wonder whether any of those who were classified as nonsmokers actually smoked under one pack a day.

As a consequence of the missing 30 responses, we are faced with a range of possible values for the proportion of smokers from 60% (73 divided by 91 plus 30) to 80% (73 divided by 91). As might be expected, the plastic surgeon used the higher figure of 80% to make the comparison.

The surgeon compared the 80% to the 33% of the general population which are reported as smokers. This seems like an extreme difference, but we have to know if this is a difference that could have occurred by chance. The statistical tool for checking such results against chance effects is called *hypothesis testing*. Simply described, you set up a "null" hypothesis against which to compare the observed result. In this case, think of the general population as consisting of tens of millions of persons, and take as the null hypothesis that our skin-sloughed face-lift patients were a random sample of 121. Then use probability theory to find the probability that such a random sample would have 80% smokers.

Without giving details of the calculation, we find that the probability that the skin-sloughed patients were just a random sample of the general population (with respect to their smoking habits) is virtually nil, about 1 in 1000, using the lowest value of 60% smokers for the skin-sloughed patients. Of course, if we had used the higher figure of 80%, the probability of this being a chance occurrence is much less than 1 in 1000.

But there is a serious methodological flaw in this analysis. The 1065 face-lift patients who were not queried regarding their smoking habits may not have had the same smoking characteristics as the general population. But it was the general population, rather than all patients, that the plastic surgeon used for his baseline. Suppose all 1065 patients had a proportion of smokers greater than 60%? Then smoking and skin slough probably would not be related. Unfortunately, the plastic surgeon did not interview the 1065 (for reasons we can understand but not forgive). This inadequate survey leaves us with suspicion of smoking as the responsible agent, but no more. It is an important suspicion and we hope that someone else will use the right tool (an experiment) properly and get a more substantive answer.

C. Is Nothing Good Enough for You?

Does this recitation of flaws in statistical experiments give you the feeling that no one will ever be able to perform an experiment which meets our demanding standards for 100% perfection? Of course not! We don't expect or demand absolute perfection, agreeing with the one of the great statisticians of all time, Jerzy Neyman:

> The tests themselves give no final verdict, but as tools help the worker who is using them to form his final decision. . . . What is of chief importance in order that a sound judgment may be formed is that the method adopted, its scope and its limitations, should be clearly understood . . . . [15]

The great majority of statistical experiments are not megamisuses and stand up to the critical investigation to which they are exposed by their publication. Many Nobel prizes have been won by investigators who have used their tools well. A large part of the remarkable improvement in public and private health is the result of sound statistical experiments. Good work *can* be done. It is a statistical crime to not try.

8
Faulty Interpretation

Easily doth the world deceive itself in things it desireth or fain would have come to pass.

Montaigne

Facts, or what a man believes to be facts, are delightful. . . . Get your facts first, and then you can distort them as much as you please.

Mark Twain

I. Introduction

Faulty interpretation is a major misuse of statistics. In this chapter we discuss faulty interpretation and some of its sources and consequences.

We count, rank, and measure to give substance to our observations. This "substance" we call numbers, which can be correctly tabulated or summarized by the use of arithmetic. But the numbers alone are not sufficient for accurate interpretation. We must know to what they refer and what they mean. *Definition* is how we decide what is counted or measured, and faulty definition is one of the factors in misinterpretation. *Conflict* is another factor in faulty interpretations. When different organizations or statisticians reach very different conclusions from the same information, the conflict itself can create faulty interpretations. The *final cause* of faulty interpretation is *blatant misinterpretation*—data are misinterpreted not through ignorance, but simply because the obvious is ignored. *Lack of knowledge* is another source of incorrect interpretation. Not everyone is aware of his lack of knowledge. Lack of knowledge can be related to the lack of knowledge of the subject matter, which we discussed in Chapter 3 (Know The Subject Matter) and is not covered in this chapter.

II. Preliminary Considerations

All statistics—both the data and measures derived from the data—are *artifacts*. The following simple examples illustrate what we mean by this.

A. How Big Is Your Apartment?

Some statistics are conceived to be discrete, such as the number of rooms in an apartment. At first glance, the number of rooms in an apartment is an integer, such as 1, 2, 3, etc. But we have all seen advertisements for 3½ room apartments, so that our concept of discrete is broadened in this case to include rational numbers.

Occasionally, because of limitations in data processing facilities or ignorance, an investigator may reduce all numbers of rooms to integers for convenience. Then rules must be made to determine the number to be assigned to each type of room. What is the number assigned to a foyer, a large closet, or a combination living and dining room? How is a "foyer" or a "combination living and dining room" *defined*? The assigned number of rooms then becomes an artifact, a consequence of the definition. Without an explanation of the definition, the resulting statistics may mislead others.

B. How Old Is Your Child?

Other statistics are conceived as continuous; that is, taking on real number values. Age is such a variable. No person has an age that is "exactly" a unique value. Age is a measurement on a time continuum, which you can measure as precisely as you wish, depending on your clock. We usually measure age in years for statistical purposes. To do this, we make arbitrary definitions about where the line is drawn to obtain integer values for age. The most common usage is age at last birthday, although sometimes age at *nearest* birthday is used. Prior to World War II the Japanese automatically aged everyone one year on the first day of the New Year, regardless of when the person was born. A baby born up to December 31 of one year automatically became two years old on the first day of the next year: "Should a child be born on December 31st, or on any day of the year prior to that date, the infant on January 1st of the year following is regarded as being two years old" [1].

Once again, the definition creates statistics as artifacts. Thus, there are no uniquely correct data!

III. Problems of Definition

A. How Poor Are Your Statistics?

Giving aid to the poor is a continuing human issue. But, in order to give aid effectively and efficiently, you must first define what is meant by "poverty."

Only then can you count the number of people who live in poverty—the poor. You must also know the change in the number of poor with time, in order to evaluate assistance and employment programs. Political reputations and human suffering rise and fall with the answers to these questions.

In the United States, a family is categorized as poor depending on the amount of cash it receives. Those receiving less than this specified amount are "below the poverty line" and counted as poor. From time to time, government officials and some members of Congress have proposed including the cash equivalent of noncash benefits (such as food stamps and medical assistance) as part of the family income.

If such a definition were used, a family could rise above the poverty line through its own sickness. If all members of a family were ill and receiving medical aid, this amount of noncash assistance, added to their regular cash income, could lift them out of the "poor" status. Thus, this family would be eliminated from the welfare rolls. But this change in their status could also mean that certain types of assistance available to the "poor" would now be unavailable to them. Thus, by practical and social standards, they could become even poorer as a result of illness! A real Catch-22, as is shown in Figure 8.1.

If the government adopted this change in the definition of poverty, the number of "poor" would drop and politicians could argue that poverty had been reduced. Even without this self-serving misuse of the revised definition, such a change would lead to underestimating the need for aid when policy decisions were being made.

This is another example showing that any statistic we compute is an artifact.

B. Would a Change in Definition Make Any Difference?

The answer depends on the subject matter and the degree of precision that you seek. We can illustrate this with the measure of unemployment used in the United States and several other countries. First, let's briefly review the origin and nature of currently used employment and unemployment statistics and see how an artifact is created.*

During the Great Depression of the 1930s, the U.S. government enacted a number of provisions to assist the destitute: the Works Progress Administration (WPA), the Civilian Conservation Corps (CCC), and so forth. The clamor was for jobs and the government was asked to create more jobs. Lawmakers and the executive branch alike were faced with the question: How many people are

*For further information see References 2 and 3 and any current issue of the monthly publication of the U.S. Bureau of Labor Statistics, *Employment and Earnings*.

Figure 8.1 A newspaper cartoonist agrees with us on the consequences of including benefits to the poor in their income. Wasserman, for the Boston Globe, copyright 1985 by Los Angeles Times Syndicate. Used by permission.

unemployed and need jobs? Only with this information could the government decide how much money and effort to allocate to job creation programs. No one had any answers, for no one knew with any reasonable degree of certainty the number of unemployed.

Consequently, the WPA embarked on a research program to define "employment" and "unemployment" and to learn how to count the number of people in each category. The result was a definition of what is now called the "labor force." The first major classification included only people involved in the money economy. It was felt that there was no point counting housewives or voluntary workers (such as people who voluntarily assist a religious institution without pay) since Congress was not considering appropriating aid for such people. The government's plan was to use separate programs to assist those who were unable to work, were too young or too old, or were not participating in the money economy.

How do you count these people? Without going into details, the reasoning was generally as follows: (1) The U.S. labor market is based on competition for jobs. (2) A person enters the labor market when that person competes for a paid job. (3) Those people who are employed have successfully competed in the labor market and thus are one part of the labor force. (4) Those who do not have jobs but are seeking them are competing for jobs in the labor market and form another part of the labor force. (5) People who do not compete for jobs, or are in institutions such as jails or hospitals and could not compete for jobs, are therefore excluded from the count of the labor force. When these concepts were first established, members of the armed services were also excluded from the labor force on the grounds that there were too few of them to matter, and as they were already ''employed'' by the U.S. government they could not compete for jobs to be underwritten by Congress.

Thus, the labor market is defined as those people who are employed plus those among the civilian, noninstitutional population who are actively seeking jobs. This sounds like a simple definition, but is it simple to put into practice? How can this information be obtained? By conducting a survey and asking people. How often do you conduct the survey? Congress and the administration wanted information monthly. Therefore, the survey was conducted monthly. What should the time reference be? If you ask someone if he or she is employed or seeking work, you must specify some precise period. The period selected was one week in a month, in particular that week which includes the 19th day. If the specified period is longer or shorter than one week, the numbers of employed and unemployed vary, but we do not go into that here. In any event, one part of the artifact being constructed is the length of the reference period.

Before the labor force artifact can be completed, a number of additional decisions must be made. Should the survey ask everyone or only those of some specified age? During the exploratory work of the 1930s, researchers realized that children under age 14 were unlikely to be either employed or unemployed. Thus the survey was initially limited to the population aged 14 and over. Subsequently it was found that very few 14- and 15-year-olds were reported as either employed or unemployed, so the lower boundary of age was raised to 16 years.

The survey designer had many more decisions to make before the artifact could be finished. For example, what about people who work on a family enterprise for which they receive no money? Should they be counted as employed or not? The decision was that if they worked 15 hours or more per week they would be defined (and thus counted) as employed.

Another question concerned the reference week. During this week some people were both employed and unemployed. In which column did they belong?

Since any one person could be counted as only one or the other, the survey designers decided that such people should be counted as employed. Determining the official "unemployment rate" involved innumerable decisions. Some of these decisions, and the resulting procedures, have been modified considerably over the years. But the basic framework of the labor force artifact remains: *The labor force is those who are employed plus those who are seeking work.* The rest are not in the labor force.

The U.S. Bureau of the Census conducts the monthly survey on a sample of the households in the United States, and the U.S. Bureau of Labor Statistics analyzes the data and publishes the reports.*

In light of the decisions which led to the present labor force artifact, we can ask: What would happen to the unemployment rate if a person who was both employed and unemployed in the reference week was classified as "unemployed" instead of "employed," as is now done? In 1984, the official unemployment rate for the population aged 16 and over was 7.5%. There were 105 million employed; included in this figure were 5.7 million who were employed but worked only part-time for economic reasons. It can be argued that they should be classified as unemployed during the time they were involuntarily not working. In addition, there were 8.5 million who were officially unemployed. If the 5.7 million who worked part-time only because full-time work was not available were added to the 8.5 million, then we would have 14.2 million unemployed, or 12.5% of the civilian work force [4, Tables 1 and 7]. Which unemployment rate do you prefer, 7.5% or 12.5%? Or something in between?

Another change has occurred: the "labor force" has been redefined. For over 30 years, until 1983, the unemployment rate has been shown as a percentage of the total *civilian* labor force. But if you add the number of *military* personnel (who are, after all, employed and working) to the civilian labor force to get a total labor force, the statistics change.

When the labor force measurement procedures were designed, the number of military personnel was proportionately smaller than it is today, and it was thought that only the civilian component of the labor force was relevant. In 1983, military personnel were added to the labor force. The result is what you would expect: The unemployment rate was immediately reduced!

For women and older men, this change made no difference, since so few are in the military. For younger men, the decrease in unemployment rate is significant. Including the military in the labor force "answers" the cries of the media and politicians that the unemployment rate among young men aged 18 to 24 is too high: When military personnel are included, the young-male unemployment rate drops. For example, the unemployment rate in 1982 for nonwhites between 18 and 24 years of age drops from 32% to 28% when military personnel

---

*An example of such a report is Reference 4.

are included. (The rate for white youth of the same age group drops only 1%, from 16% to 15%, because relatively fewer of them serve in the military forces [5, Table 3].)

Even these (reduced) rates are too high; nothing was accomplished by changing the definition of the labor force except a change in the interpretation. Unfortunately, however, this "paper" reduction in unemployment rates could be touted by politicians as an improvement and result in a drop in federal allocations for programs for the unemployed.

IV. Problems of Conflicting Numbers

A. The Case of India and Its Food Supply: Who Knows?

Is India able to grow sufficient food for its own population? Yes, according to M. T. Kaufman in an article in the *New York Times* [6]. The article begins: "It has been four years since the last shipload of foreign wheat arrived in India."

He then continues with information based on a World Bank analysis and Indian government statistics: "World Bank analysts point out that as grain production has increased, and prices have dropped, consumption has grown to approach minimal levels of nutrition [6]." The increased production of grain (or all foods, the article is not completely clear here) reportedly came about from the introduction of new agricultural methods, the maintenance of stable farm prices, and other governmental measures.

We ask: How minimal is minimal? According to the World Bank, in 1980, the daily calorie supply per person in India was 1880 calories, only 87% of the daily requirement. The daily calorie supply per person of 1880 calories is an average, which means that many people consumed more than 1880 calories and many consumed less. We suspect that the average is 1880 calories because the fortunate few who eat a great deal more than 1880 calories offset the larger number who consume less than 1880 calories. Thus, there could be a large number of people who do not approach 1880 calories per day, which is still only 87% of the daily requirement. If you compare the U.S. average calorie consumption of 3860 with the Indian figure of 1880, you see a shocking difference—the average Indian caloric intake was less than half of the U.S. average [7, Table 24].

What was the average consumption of calories per person before 1980? Information for 1960 leads us to believe that there has either been no change over the last 20 years or a small decrease. In 1960, the calories consumed in India per person per day was reported as 1990 [8, Table 1240], which is only 6% higher than was reported for the year 1980. Because of the likely error inherent in the average calorie consumption data, we are inclined to take "no change" as the verdict.

Let's look further into the available information on food production in India. The United Nations Food and Agricultural Organization (UNFAO) gives data [9] which conflict with the *New York Times* story [6]. The indexes of food production per person in Table 8.1 show year-to-year fluctuations but no steady improvement in the food supply.

Who is right, the World Bank or the UNFAO? To try to answer this question, we look more closely at the statistics. If the World Bank's numbers refer to total food production, then they are obviously different from the per capita figures cited by the UNFAO. If the World Bank numbers refer only to grain, then grain production could have increased while total food production per person did not. This would have occurred if production of nongrain foods was cut back at the same time, which is not an unlikely event, since resources such as land are often diverted from one agricultural crop to another. Another problem is that UNFAO sometimes publishes slightly different values for index numbers for the same reported year. For example, the per capita food production index for 1979 was reported as 97 in the 1981 volume and as 101 in the 1983 volume [9]. We assume that the latest values are likely to be more nearly correct and have used the most recent values in our discussion.

Further, statistics on agricultural production, regardless of the agency which publishes them, often contain errors of unknown size. An example is given in the next section. Another example of highly questionable agricultural production statistics from Ethiopia, described by William Abraham, is shown in Part V of Chapter 13.

Can we say any more, based on our examination of these reported statistics, than that India is just holding even? If India's ability to feed itself is an important issue, then don't these two prestigious organizations owe us a recon-

Table 8.1 Indexes of Indian Food Production (per person)

Year	Index
1969–1971	100
1976	98
1977	107
1978	108
1979	101
1980	102
1981	108
1982	102
1983	114

Table 8.2 Conflicting Estimates of Coffee Production in El Salvador

Source of estimate	1980–1981 coffee production (tons)
Salvadoran government	25,950
U.S. government	20,204
University of Central America	16,951

ciliation of their discrepancies? And shouldn't the author of the article which relayed this information have pointed out that alternative and apparently contradictory information exists? Shouldn't the report also have defined whether caloric consumption included only grain or *all* foods? In the end, all the reader can do is to *stay on guard* and not automatically believe all the published interpretations.

B. Agricultural Production in El Salvador

The effect of land reform in El Salvador was an important political issue in 1983. Illustrating an excellent example of investigative reporting, R. Bonner of the *New York Times*, reported on agricultural production and presented all the information available [10]. Three concerned groups, the Salvadoran government, a U.S. Government agency (the Agency for International Development), and the University of Central America (Center for Documentation and Information), made statistical estimates of production and its change over time. The statistics on 1980–81 coffee production, shown in Table 8.2, are a good example of how the production estimates can differ.

The differences speak for themselves. But what about the reporting of these estimates? When three agencies differ so much, can we give credence to estimates which attempt to give an impression of exactitude by reporting numbers such as 20,204 or 16,951? In our view, the only reasonable reports are 20,000 and 17,000.

V. Blatant Misinterpretation

At the start of this chapter, we defined blatant misinterpretation as the result of the reporter or researcher ignoring the obvious—what the dictionary calls "glaringly evident" [11, p. 97]. Not everyone sees the obvious until it is pointed out, when it becomes glaringly evident. This is why all of us have made such misinterpretations at one time or another, and it is through the open exchange of information (as in Letters to the Editor and scholarly commentary) that we slowly reduce the number of occurrences of blatant misinterpretation. You will need only logical reasoning to follow our first example (Section A), but the second example (Section B) involves both logical reasoning and some elementary idea of a statistical analysis concept—probability—which we give you.

A. Making Sense Out of Numbers

1. The Quality of Teeth

In an article in the *New York Statistician*, Mark L. Trencher brought to our attention a statement by *U.S. News & World Report* on tooth decay:

> . . . the rate of tooth decay among school children has dropped roughly one-third in 10 years, indicates a recent survey by the National Institute of Dental Research. This nationwide sampling . . . showed that 37 percent . . . had no decay. A comparable sampling 10 years ago showed that only 28 out of 100 children were free of decay—a difference of 32 percent from the current population. [12]

The "32 percent" is the reduction in percentage points of the number *without* cavities obtained by dividing the reduction in percentage points (37% minus 28%) by 28%. But the article is about the reduction of those *with* decay!

The reduction in the rate of those having tooth decay is only 12.5%. To get this value, first you find that the percentage having tooth decay 10 years ago was 72% (100% minus 28%) and that it is now 63% (100% minus 37%). The reduction in the rate for those having tooth decay is thus 9 percentage points (72% minus 63%). Divide this reduction in percentage points of those having decay by the rate for 10 years ago and you get 12.5% (9% divided by 72%) as the percentage reduction of those having tooth decay. A reduction of 12.5% in the rate of tooth decay among school children is a cause for joy and encouragement, but it is *not* a reduction of "roughly one-third."

2. There Were Plenty of Jobs for the Unemployed—Or Were There?

At a news conference in 1982, President Reagan suggested that unemployed people should look at the help-wanted sections of their newspapers, where they could find many opportunities. The President said that (in the January 17, 1982 issue of the *Washington Post*): ". . . there were 24 full pages of classified ads of employers looking for employees."

However, according to the AFL-CIO's *Revenews*:

> Actually there were 26 pages of help-wanted ads.* . . . But of these 26 pages, four

*This is one of those rare cases where an error in basic data was in not in a direction supporting the speaker's contention (in this case, of a plentitude of opportunities). This supports the view we express in Chapter 1 that you should be cautious in attributing misuses or bad data to deliberate actions. The fact that most such errors tend to support the speaker's or author's views does not mean that they deliberately committed these errors. We all make errors.

> were devoted entirely to the engineering profession, three to nursing, three to computer programming, two to secretarial work, and two to accounting and bookkeeping. . . . [A total of] no more than 3,500 job openings, but there are now more than 85,000 unemployed people in the Washington Metropolitan Area. [13]

Twenty-four (or 26) pages of classified ads for help may seem ample. But the total number of opportunitites is small compared to the number of the unemployed, and most of the opportunities are for skills which many of the unemployed don't have!

B. Choose Your Drink

Sometimes you need statistical analysis (but not necessarily a great deal) to verify whether or not an interpretation is correct. In 1981, the Joseph Schlitz Brewery Company staged live taste tests of its beer against competing brands during the Super Bowl and two National Football League playoff game telecasts. In each of these three cases, 100 self-identified regular users of competing brands of beer (Miller, Budweiser, Michelob) were given blind, paired comparison preference tests with respect to Schlitz's beer.

In a "paired comparison preference test" the test subject compares two products and states a preference. A test is "blind" when the test subjects do not know which of two products they are testing when they make their choice. It is well known among market researchers that a fundamental problem with such tests is that, occasionally, some test subjects have expressed a preference for one product over another when the two products were identical [14]!

If the two products are definitely different, such as steak and carrots, a preference may be taken as real. But if a paired comparison of two *similar* products results in an even split of preferences, we have no way of knowing whether this is because of true preferences or the inability of the test subjects to taste a difference between the two products. They may simply be guessing.

The numerical results were as shown in Table 8.3. In reporting the results, the announcer emphasized the high percentage of testers (ranging from 37% to 50%) who preferred Schlitz to their usual brand. Unfortunately for the Joseph

Table 8.3 Results of the Schlitz Taste Test

Competing brand	Preferred Schlitz (%)
Miller	37
Budweiser	48
Michelob	50

Schlitz Brewery Company, the message may not be that beer drinkers prefer Schlitz's beer to their regular brand, but rather that beer drinkers cannot tell one brand from another when they make a blind paired comparison. This is an important message for those concerned with the marketing of beer, but not the same as the message that the brewery wanted its television viewers to hear when it announced these results.

To evaluate these results, we hypothesize that the results come from the inability of the test subjects to tell the beers apart, and we judge the results by their deviation from this hypothesis.

If beer drinkers can't discriminate between beers by taste, then we expect that blind paired comparisons will show 50% of the test subjects (or 50 out of 100) stating a preference for Schlitz.

Using this hypothesis, we can compute the probability of getting the observed results. However, you do not have to perform any computations to see that if the drinkers were unable to tell the difference, the results for Budweiser and Michelob are likely (48% and 50%, respectively, expressing a preference for Schlitz). To evaluate the result for the comparison with Miller, you can easily compute the probability of occurrence to be less than 1% (and verify our results using methods discussed in Chapter 10).

The statistics give no reason to believe that the Michelob and Budweiser results are other than guesses, whereas there is strong evidence that the majority of Miller drinkers prefer Miller. This is a different interpretation than Schlitz reported. For the television viewer, the reasonable conclusions based on this test are: (1) Those who are regular drinkers of Budweiser and Michelob can't taste a difference from Schlitz; (2) Those who are regular drinkers of Miller can taste a difference between it and Schlitz; and (3) Most regular drinkers of Miller prefer Miller, but about one-third prefer Schlitz by taste.

As far as we can tell, this test was well-designed and executed, but we regard its interpretation as a misuse. In a controlled situation, a market researcher could make excellent use of this test by digging deeper: Who were the individual Miller drinkers who preferred Schlitz and how did they differ from the other Miller drinkers? Who were the Michelob and Budweiser drinkers who chose Schlitz in the comparison? Could they tell a difference, or were they in fact unable to discriminate? If they preferred a taste difference, why? If they were unable to discriminate, why? Those market researchers who use (as most do) good basic data and sound statistical methodology, paired with correct and careful interpretation, can make better product decisions.

C. One Giant Logical Leap for Mankind—Unfortunately, a Misstep

Dr. Stephen Kellert (of Yale University) and Dr. Alan Felthaus (of C. F. Menninger Memorial Hospital) examined the association between childhood cruelty toward animals and aggressive behavior in adult criminals and noncriminals, noting that: "The research literature appears to suggest, thus, that childhood cruelty toward animals may operate as part of a behavioral spectrum which is associated with violence and criminality in adolescence and adulthood" [15, pp. 5–6].

Three groups of males were examined: aggressive criminals, nonaggressive criminals, and noncriminals. The criminal populations were drawn from federal penitentiaries and noncriminals were chosen at random from areas near the penitentiaries. Childhood behavior of these individuals toward animals was determined through interviews with the test subjects, parents, and siblings.

There are many methodological questions to be raised in the conduct of this study, which the authors themselves discussed and either dealt with or took into account in their analysis. Their methodology is not our concern at this time, and we will assume the validity of their conclusion:

> This paper has reviewed a number of results from a study of childhood cruelty toward animals, motivation for animal cruelty, and family violence. The strength of these findings suggests that aggression among adult criminals may be strongly correlated with a history of family abuse and childhood cruelty toward animals. . . . This [sic] data should alert researchers, clinicians, and societal leaders to the importance of childhood animal cruelty as a potential indicator of disturbed family relationships and future anti-societal and aggressive behavior. [15, p. 33]

In short, a child who displays cruelty toward animals and becomes a criminal is more likely to be an aggressive criminal or antisocial than a child who is not cruel to animals. What action might be taken as a consequence? As the authors suggest, the behavior of children toward animals might be used as an indication of future problems. That much, and no more, is a legitimate claim, as derived from their investigation and analysis.

But the authors now make a giant, and invalid, logical leap: "The evolution of a more gentle and benign relationship in human society might, thus, be enhanced by our promotion of a more positive and nurturing ethic between children and animals" [15, p. 33]. The authors have shown that children who were cruel to animals are more likely to be aggressive adult criminals, but they have *not* shown that the promotion of kindness to animals in a child will result in that child being less likely to become an adult criminal. Association is not cause (as we discussed in Chapter 7); although it may well be that there is an underlying causal factor—the authors themselves mention several—which causes both childhood cruelty to animals and adult aggressive criminality.

To be able to say that the "promotion of a more positive and nurturing ethic between children and animals" will reduce aggression in adult criminals, you must run a controlled experiment (as discussed in Chapter 7) in which two or more groups of similar children are exposed to "promotion" and "nonpromotion'' of kindness to animals and then followed through their adult lives. (Most researchers use identical twins in studies such as this.)

For now, the jury is still out, despite these researchers' good work and intentions.

VI. Summary

Misinterpretation arises from many causes, some of which are discussed in this chapter. Cause is less important than the ability to avoid its consequences.

Some guidelines for readers (and writers):

Headlines. Don't take headlines seriously. They are written to catch the reader's attention, and are not necessarily accurate.

What are the definitions? Definitions by which data are collected or tabulated may not lead to answers to the questions being asked. Always try to find out the definition used to categorize, which good writers and authors will give in the text of their articles and reports.

Changing times. Definitions may change with time. Determine if the definition has changed.

Know the subject. Knowledge of the subject is relevant to collecting data and drawing conclusions. To avoid being misled by a misuse of statistics, know the subject or consult an expert.

Baseline values. Incorrect baseline values lead to misleading results in comparisons, as for example, in the discussion concerning tooth decay in Section A, Part 1, of this chapter. Find out what is the base and if it is consistent. You can usually determine the base by a simple computation, as we have done in this chapter, and elsewhere in this book.

Spurious precision. Since spurious precision gives false confidence in interpretations, round published statistics to numbers that make sense in view of the way in which the data were collected and processed.

9
Surveys and Polls: Getting the Data

I am not bound to please you with my answer.

Shakespeare

It is better to know nothing than to know what ain't so.

Josh Billings

I. Introduction

You survey to know the general state of affairs: opinions, events, attitudes, intentions, ownership, habits, purchases, demographic characteristics, and so forth. Observation provides the basic data for a survey. You count cars passing an intersection, report about the condition of bridges, and ask questions of consumers, voters, and inhabitants. Surveys have an ancient history and their types and purposes are as diverse as the range of human activity.

There are many types of surveys, but today those in which we pose questions to people and then record, analyze, and interpret their responses have the most impact in the political arena. Because this type of survey is so important, it is the only type we discuss, and in this chapter we give examples of misuses of statistics in surveys of this type. We cannot give examples of all possible misuses, for they would fill several volumes. However, we do illustrate the kinds of reasoning which you can use to recognize and analyze these (and other) misuses of surveys.

A survey has three basic parts: the questions, the responses, and the analysis. The final product—the report—often appears scientifically correct whether it is or not. Professional pollsters and surveyors are aware of the pitfalls inherent in this process and usually try to avoid them, but even the professionals don't always succeed. Unprofessional pollsters and surveyors are often unaware of the pitfalls and can make so many errors in polling or surveying that much of their work qualifies as a megamisuse.

II. First Principles

A. To Census or To Sample?

A survey may include all of the population of interest or only a part of it. Statisticians call the entire population of interest the *population* or the *universe* and a part of the population a *sample*. If the survey includes all of the universe, it is called a *census*. If it includes only a part of the universe, it is called a *sample*. Some surveys include all of the population for one set of purposes, together with a sample of the same population for other purposes, as in the case of the 1980 U.S. Census of Population and Housing (which we discuss later).

Whether we take a census or a sample, we pick a group of individuals and ask them questions. Some of them will respond (and become *respondents*) and some will not (and become *nonrespondents*).

The U.S. Constitution requires a count of the population every ten years to allocate members by states to the House of Representatives. Legislation adds the "one person, one vote" rule, which requires that all congressional districts within a state be as equal in population as is possible. The 1980 U.S. Census of Population must first satisfy the state by state counting requirement of the U.S. Constitution. Then, to satisfy the "one person, one vote" requirement, the Census must get a complete count of everyone in every city, county, and even smaller defined areas. For these purposes, a complete, nationwide count is conducted every ten years. The 1980 U.S. Census asked only six questions of every household and its members.

There is no Constitutional requirement to collect information on other subjects, and where other information is sought, a sample generally is used. In the 1980 U.S. Census of the Population, the additional data collected were responses to about 56 questions. "About" 56 questions? Yes, the number of questions is approximate because some questions had several parts and the number depends on what parts are counted as separate questions.

The *sample* questions were asked of every sixth housing unit (about 17%) in most of the country. If an administrative unit—county, incorporated region, or other local division—contained fewer than 2500 people, the sample questions were asked of every other housing unit (about 50%) [1, Appendix B, p. B-2]. The answers from these samples are projected to the population of each sampled area by using the ratio of the sample to the counted population of the area, a process which is subject to sampling error (as is discussed in the next chapter).

A supplementary sample of the Native American population also was taken in 1980. All Native Americans (American Indians, Eskimos, Aleuts) living on Federal or state reservations, or the historic areas of Oklahoma, were asked additional questions. Native Americans living elsewhere were not included in this supplementary sample.

B. How Big Is Big?

The size of surveys varies greatly, from a dozen respondents to millions, as in national censuses of large countries such as China and India. Big or small, they all have problems to resolve.

Large surveys (both censuses and samples) involve large numbers of interviewers, which creates problems in training, supervision, and the processing of great amounts of data. In a country such as India, for example, where many languages are spoken, interviewers must be found who can converse fluently in several languages.

Small surveys are almost always samples. Because they seem so easy to carry out, market researchers and pollsters use them frequently, as most of us know—we get mail questionnaires and phone questionnaires all the time. But no matter how small the sample, if the survey is to avoid being a misuse, it must be designed and carried out properly. Unfortunately, we see many misuses. When a television reporter sets up a camera in the street and casually interviews passersby, the reporter is conducting a survey and the audience may incorrectly assume that a scientific process is taking place and take the responses for accurate and precise determinations.

C. I Do Not Choose to Answer

If the *response rate*—the proportion of the sampled group which responds—is reported with the results, you can draw your own conclusions as to the value of the findings. Unfortunately, response rate may be confused with the proportion of the total universe which is sampled (the *sampling proportion*).

We have seen surveys in which the response rate was as high as 95% and others in which it was as low as 3%. A low *response rate* is a problem because the nonrespondents may not respond for a reason that is related to the subject of the survey, making the results suspect (we discuss what is ''low'' later). On the other hand, a small *sampling proportion* is usually of no significance. As Dr. George Gallup (the famous pollster who has released over 6000 polls in over four decades of polling) points out, you do not need a large sampling proportion to do a good job, if you first stir the pot well:

> Whether you poll the United States or New York State or Baton Rouge (160,000 population), you need only the same number of interviews or samples. It's no mystery really—if a cook has two pots of soup on the stove, one far larger than the

other, and thoroughly stirs them both, he doesn't have to take more spoonfuls from one than the other to sample the taste accurately. [2, p. 11]

D. How Many Questions Should I Ask?

The number of questions asked in surveys varies from one to hundreds. A voting intention survey may ask no more than "Do you intend to vote?" But in an "in-depth" survey hundreds of questions are asked. The Archdiocese of New York surveyed the Hispanic population in its area to investigate twelve subjects of interest, including "Religious Identity," "Meaning of the Church," "Religious Education," and so forth. This in-depth survey contained 400 questions [3, p. 105 ff.]. Customer product preference surveys fall between these extremes, asking for information concerning individual demographics (age, sex, family income, and so forth) and product preference (which product do you prefer and why?).

III. Examples of Survey Misuse

A. Verification: What Did They Say?

You can't verify many of the surveys in which people are asked questions. In most cases, the best that you can do is make a spot check of some of the answers. This is what the U.S. Bureau of the Census does when it resurveys a sample of the original respondents for verification. We get some verification, also, when two polling organizations get substantially the same results from independent surveys.

Strict verifiability means that you must repeat the survey in the same way under the same conditions. For the kinds of surveys we are discussing, it is almost impossible to have the same conditions repeated and hence, strict verifiability is almost always impossible.

Even if you could reproduce identical conditions, repetition alone can change the individual's response. The U.S. Bureau of the Census data on reported length of unemployment is an example of this phenomenon. When making the monthly sample survey of employment and unemployment (Current Population Survey, or CPS), the Bureau's interviewers question the same housing unit for four consecutive months and usually are able to reinterview the same persons. The interviewers ask those who say they are unemployed to give the number of weeks of unemployment, and this question is repeated as long as the same respondents say they are unemployed.

Since the time interval between interviews is known (four or five weeks) you would expect that the number of weeks of unemployment reported at the second interview would be greater than the first value reported by the number of weeks between interviews. Alas, only about one-quarter of the respondents give consistent responses! Some "gain" weeks of unemployment between interviews, others "lose" weeks of unemployment [4]. Which response is correct: the first, the second, the third, or none of them?

Economic factors often make verification impractical. One of us directed a survey of the employment of physically handicapped workers under a grant from the U.S. Office of Vocational Rehabilitation [5]. After publication, another researcher wanted to investigate the same subject to confirm or refute the results. Since this survey requires a sizable staff, he tried to obtain grant funding, without success. There was no verification.

Human attitudes and social environments change with time. Thus the passage of time alone can prevent verification. If the employment researcher of the preceding paragraph were to obtain funds today, he could no longer verify the physically handicapped worker survey of 1959. He could only perform another survey which "duplicated" the 1959 survey in its questions and methodology. The outcome would be the basis for estimating the changes which have taken place since then, but strict verification would be impossible.

Some verification is better than none. We can get limited verification when several investigators survey the same population on the same topic at about the same time. Five polls of public reaction to the U.S. presidential candidates were made by separate polling organizations in July 1984, and the *New York Times* published the results [6]. The questions and survey procedures varied among the polls, but the purpose of all five was the same: to find what proportion of the electorate favored Reagan, Mondale, or neither. The reported proportions favoring Reagan varied from 44% to 60%. These extremes are far outside our estimate of the variability which can be attributed to sampling error alone.*

Thus, there is no verification of the proportion favoring Reagan. However, there *is* verification of the result that Reagan was ahead in voter preference since all five polls agreed on that point.

Lack of verifiability means that the reader often must accept or reject results based on the reader's confidence in the surveyor.

Reputation counts. In the absence of a definitive way to verify survey results, your best defense is a thorough awareness of survey methodology, so that you can decide how to interpret the results of a particular survey.

B. Bias in the Questions Asked

How many are the ways to bias the questions asked of respondents! The bias may, or may not, be deliberate. There are many ways in which the same question can be asked and answered, and it is not certain that all pollsters agree on the best

*We compute a sampling error of plus or minus 2% using techniques which are described in the following chapter. If you master the simple methods shown there to estimate sampling error, you will be able to make such computations yourself; they do not require mathematical knowledge.

way to ask questions. Also, professional pollsters often disagree on which way of asking is least likely to be biased.

There probably is no way to eliminate all bias from all questions in all surveys or even from some questions in some surveys. Here we discuss only two types of bias: *leading questions* and *predetermined answer categories*. Through these two examples, we illustrate the type of reasoning needed to deal with bias. Awareness of the nature and effect of bias is your best defense.*

1. Leading Questions—Getting the Answer You Want

When a question is worded so as to ensure getting a particular answer from the respondent, it is called a *leading question*. There is no shortage of examples of leading questions. In 1975, one of us received a questionnaire soliciting support for opposition to H.R. 77 from William L. Dickinson, then a member of the House of Representatives. The second question of this questionnaire was:

> 2. Do you feel that anyone should be forced to pay a union boss for permission to earn a living?
>
> YES _______ NO _______

It is extremely unlikely that anyone would answer "YES" to this question, for who wants to *force* someone to pay a union *boss* for permission to earn a living? But some people who do not want workers to be forced to pay union bosses for permission to work may think closed shops are good or that not all union leaders are bosses.

In 1981, Guy Vander Jagt, then a member of the House of Representatives, distributed a "questionnaire" for a "National Voters' Survey on President Reagan's Economic Recovery Plan." The first "question" was:

> 1. Federal Tax Cuts. The Reagan Administration bill calls for a permanent 30% across the board reduction (10% a year for three years) in your individual federal taxes in order to stimulate savings, investment and create more jobs.
>
> I urge Congress to vote _____ for this proposal.
> _____ against this proposal.

This "question" will produce a response in favor of the questioner's point of view, as in the prior example. Almost all Americans want "to stimulate savings, investment and create more jobs" and to reduce their taxes.

*Two books that deal with bias in detail are cited in References 7 and 8.

We don't think this is really a survey question ("I urge Congress to . . ."). Congressman Vander Jagt merely called it a survey and then counted the responses. We see many "surveys" of this type.

In addition to leading questions in these two questionnaires, the originators asked for contributions of money to support their proposals. In the first example, funds are solicited for the National Right to Work Committee and in the second case for the 1981 GOP Victory Fund. This is a misuse of a statistical tool, since the purpose of a questionnaire should be to solicit data and only data. The request for funds creates a bias, because the request itself clearly signals the viewpoint of the surveyor. Those who do not support the proposal are unlikely to respond at all. Thus, the surveyor ends up with an overwhelming vote of support, to be quoted to the press and in Congress, but not necessarily reflecting the congressman's constituency.

Rene Levesque, while he was Premier of Quebec, wanted more self-determination for his province from the federal Canadian Government. He sought support by referendum, which is a form of survey. A cartoonist (undoubtedly one who had received many questionnaires) suggested three answer categories for the referendum, which would assure that the Premier got the answer desired [9]:

Which would you prefer?

A—No more hockey.
B—No more sex.
C—Sovereignty Association.

2. Predetermined Answer Categories—Keeping the Respondents from Giving the Answers They Want

In many surveys, the respondent is given a list of answer categories from which to choose an answer. Usually, the respondent can't give any answer other than one which is specified by the given category.

The organization Zero Population Growth ran a survey using a questionnaire in which 12 questions were asked. Some of the questions were:

Do you support sex education programs in schools?

Do you favor a national policy to plan for stabilizing population size?

Do you favor increased development aid or trade agreements with foreign countries in order to reduce illegal immigration into the United States?

Respondents could only give two answers, "Yes," or "No," for all the questions. There was no category for "Don't Know," "No Opinion," or "Can't Decide." Thus the respondent whose opinions fell into any of these other categories had no choice other than not answering. Any surveyor can avoid this misuse of statistics by including "Don't Know" categories in the questionnaire.

In 1983, U.S. Senator Bill Bradley sent a questionnaire to residents of New Jersey asking for "guidance in considering the problems of the Social Security System." He asked the recipients to "rank in order of preference the three solutions (they) prefer." He gave nine alternatives from which the three were to be chosen. What are respondents to do if they think the best solution is not one of those on the list of nine? They are forced to either choose among the nine or fail to respond. It is better to offer the respondents an "Other" category and allow them to write other solutions where they feel it's appropriate. But will they take the trouble? Perhaps they will only do so if they feel strongly about a particular issue. As it stands, the Senator cannot be sure that the answers to his questionnaire truly reflect the opinions of the residents of New Jersey.

3. *Interpretation—What Does It All Mean?*

How questions are asked—and answered—influences the interpreted results. The results quoted in the final presentation (which is often a newspaper article on surveys affecting public policy) may or may not truly reflect the opinions or intentions of the respondents. To evaluate survey results, you must know exactly how a question was worded and how the sample was interviewed. In Figure 9.1 you will see how artist George Mabie views this important issue.

The need to know exactly how a question is worded becomes crystal clear when the same information is requested by two differently-worded questions. The New York Times/CBS News Poll ran an experiment which clearly illustrated how asking for the same information with two different questions can give results which lead to different interpretations [10].

In this experimental poll, respondents were first asked, "Do you think there should be an amendment to the Constitution prohibiting abortions, or shouldn't there be such an amendment?" Sixty-two percent of those questioned opposed the amendment and only 29% favored it.

The same people were later asked "Do you believe there should be an amendment to the Constitution protecting the life of the unborn child, or shouldn't there be such an amendment?" Thirty-nine percent were opposed to the amendment as described in this question and 50% favored it. When the issue was phrased without reference to a Constitutional amendment, 62% said that a woman should be allowed to have an abortion and 15% said that it depended on circumstances.

E. J. Dionne, in discussing this experiment, interprets the results with respect to answers to other questions on this subject [10]. His conclusion, which is more complex than we can fully describe here, stated that, generally, more people favor abortion than oppose it. But if you try to decide whether the population of the United States was for or against abortion simply on the basis of the first two questions, it is difficult to arrive at a clear-cut answer. A headline

"And now, Sirs, in your work—which do you find most efficacious—soap flakes or scouring powder?"

Figure 9.1 Some problems of how questions are worded and how samples are interviewed have been with us for a long time. The artist based this drawing on his observations of polling practices in the 1930s. (Reproduced by permission of the estate of George Mabie.)

based on only one of these two questions, saying either "U.S. Favors Abortion" or "U.S. Opposes Abortion" is a misuse of statistics.

C. Sampling Done Here

We take samples because we cannot afford the money, time, or resources needed to query the universe. We want to project the results of the survey of the sample to its universe; but how many pitfalls there are on the way!

1. *A Depressing Survey*

The year is 1936 and the surveyor wants to survey the universe of all voters. The size of this universe is tens of millions, so a sample is taken. But from what listing of people are we to take our sample? It is easy to get lists of automobile registrants, telephone directories, and similar sources. But this is a time when automobile and telephone ownership is a luxury. Thus the sampled universe is not the target universe (of all voters) and is of a higher economic and educational status. We should not be surprised when the sample is wrong about the actions of the voters, as was the survey based on the sample just described, taken by the *Literary Digest* in 1936 [11]. The magazine said that Alfred M. Landon would win the election. Instead, Franklin D. Roosevelt was elected with 61% of the popular votes cast.

2. *The Self-Chosen People*

The surveyor can take extensive precautions to assure that the sampled universe is the same as the universe of interest, but if it is possible for the recipient of a questionnaire not to respond, the results can be seriously distorted. For example, the *Literary Digest* survey used mail questionnaires. Only 20% of the recipients responded—the *self-chosen.*

> There is general agreement that this mail ballot method was subject to a serious distortion because the better educated and more literate part of the population, as well as those who were higher on the economic scale, tended to return their ballots in greater proportion than those who were lower in educational and economic status. [11]

A more recent example is a telephone survey to determine voters' opinions in the 1980 U.S. presidential election. Listeners to the Carter-Reagan debate televised on the ABC network were asked to say who they felt won the debates. All of the sample of listeners was self-chosen, since it was necessary first to listen to the debates and then to make a 50¢ phone call in order to become a respondent [12]. The response was 2:1 in favor of Reagan, and this is how the story was reported. Since Reagan won the election by 1.16:1, the method was not criticized as was that of the *Literary Digest* poll. But the survey was no more valid than the 1936 survey and many similar self-chosen surveys which mislead the public and policy-makers.

Many of the poll results appearing in magazines and newspapers fall into this category. *Playboy*'s 1982 poll on sexual attitudes resulted in headlines projecting the sexual attitudes of the self-chosen to the universe of all inhabitants of the United States [13].

News media continually give us new forms of the old misuses. The two-way connection possible with certain cable TV installations has led to the use of "instant polls" in which viewers can, *if they choose*, push buttons and answer questions presented on the video screen. The results are tabulated within milliseconds and summarized on the same TV screens. There is a hierarchy of self-selection: allowing or requesting the installation of the cable TV, watching the particular program, choosing to respond.

There is nothing inherently wrong in a self-selected sample. However, it is wrong when a Dr. Nowall projects the results of a survey of such a sample to a universe that is not represented by the sample. For example, the ABC network "made no claim that the method of polling would be a good predictor, and during the following day, indicated that there was some controversy as to the scientific accuracy of the method" [12]. It would have been correct to say that, of those who watched the Carter-Reagan debates and chose to spend 50¢ to respond, the opinion was 2:1 in favor of Reagan. This is a valid statement, but to project this result to all listeners or to all voters is a misuse of statistics.

The same statement can be made in regard to the *Playboy* poll or any other similar poll relying on the self-chosen. Our many examples show that we cannot always rely on the public, headline writers, or policy-makers to pay close attention to "scientific" qualifications which would limit the conclusions only to the sample. Public opinion and national policy are frequently influenced by invalid projections of such sample survey results to the whole population.

For example, here is a congressman relying on self-chosen respondents for aid in making legislative decisions:

> Congressman Hollenbeck has received from the survey an extremely valuable information source to aid in making legislative decisions best representing the views and concerns of (N.J.) Ninth Congressional District residents. [14]

Each year, Congressman Hollenbeck mailed questionnaires to voters in his district and requested replies. In the report cited above, he notes that he received 16,000 replies. Because the respondents were only those of his constituents who felt strongly enough about some issues to respond, and took the time to respond and to mail back the questionnaires, the results cannot be statistically projected to the universe of all constituents as might be done with a random sample.

3. *The Good and the Bad: Telephone Interviewing*

In the beginning, interviewers sought out respondents and dealt with them face-to-face. If the sample of respondents was scientifically chosen, if all who were approached responded, if the questions were well-phrased and the interviewers were trained and skilled, the quality of data collection was assured.

Later, someone had the idea of sending questionnaires by mail and questionnaires were sent to the four winds even as pollen is distributed by air currents. The analogy is not so far-fetched since, like pollen, not all the questionnaires found anyone home and not all which settled down received attention, making the problem of the self-chosen respondent important.

Today, we have the telephone interview. Surveyors have found that the telephone interview is less expensive even when, in a few cases, the surveyor must send an interviewer for a face-to-face interview. Studies have shown that the results of telephone interviewing differ little from face-to-face interviewing for some subject matter [15]. In such cases, any decisions based on the survey findings would be essentially the same no matter how the interviewing was done.

However, many poor persons—often members of a minority group—have no telephone in their homes. In 1983, close to one in ten of all families had no phones, and two in ten of those with annual family incomes below $10,000 had no phones. More than three in ten of black and Spanish-origin families with annual family incomes below $10,000 had no phones. These people cannot be reached by a telephone survey. Suppose this is a survey of urban housing problems. Can you project the results to include those who live in low-income housing?

The purpose of the survey should determine the accuracy of a telephone sample, because it is the survey's purpose which delineates the target universe. If the purpose of the survey calls for a target universe of middle and upper income families, there is no problem with a telephone survey. All those families have phones. But in a survey such as the Archdiocese of New York survey of the Hispanic population [3], the data from a telephone survey could have been disastrously flawed.

4. *Not Everyone Will Talk to You*

A sample may be designed to be appropriately random so that probabilistic projections could be made, but design is one thing, implementation another. The nonrespondent is the major problem in implementation. If, for example, only 10% of the chosen sample responds, probabilistic analysis cannot be used (standard errors are without meaning) and the results may be totally useless. Why do some people refuse to be interviewed? If the reasons for refusing to be interviewed are linked to some characteristic of the survey questionnaire, then the results may be biased. Thus, in the implementation of the survey, every effort must be made to get 100% response rates from the chosen sample.

What response rate is adequate? Anything less than 100% is suspect. Good surveyors aggressively and skillfully try to get 100% response. If the interviewing is done face-to-face or by telephone, it may be possible to make return visits or calls. The problem of obtaining responses is more complicated in mail surveys. Professional surveyors use many procedures for dealing with the failure to respond. Without going into the details of these procedures, we advise readers to be wary of mail surveys. And if the report doesn't tell what proportion of the sample responded, *beware*.

Even when the proportion is reported, it may be unsatisfactory. A survey to determine attitudes of both business people and academics toward several aspects of Master of Business Administration (MBA) degree programs included both presidents and personnel directors of many companies. The results were that:

> . . . business participation was low, even though the presidents' and personnel directors' response rates fell within the 20% to 25% range that is standard for mail surveys. Readers should be aware, therefore, that the business response is not as representative as we had hoped it would be. [16]

The "20% to 25%" who responded were self-chosen. The disclaimer quoted above is a mild caution where a drastic warning is needed because the results of this survey (based on a response that represents the opinions of a self-chosen minority) could be used to determine educational policy for MBA programs.

Cost limits the amount of follow-up and, hence, the return rate. Usually, even with intensive efforts, it is almost impossible to get 100% response (deaths, disappearances, physical incapacity). If we have reason to believe that the nonrespondents are a random sample drawn from the same universe as the respondents, then both groups have the same characteristics, subject to chance effects. Is there any way to come to this conclusion at some reasonable level of confidence? One way is to take a modest sized random sample from the nonrespondents and then concentrate extraordinary efforts on getting responses. This "survey within a survey" can be used to make confidence intervals on the characteristics of the nonrespondents.

In the absence of 100% response, or information on the nonrespondents, the probabilistic computations (confidence intervals, levels of significance) are in error. The larger the proportion of nonrespondents, the worse the error. Even at a nonresponse level of 50%, probabilistic statements are likely to be in serious error.

To the best of our knowledge, a response rate of 100% is rarely achieved. Then what response is acceptable? It is not possible to give an exact value for the minimum acceptable response rate and to reject any survey with a response rate below that value. We can say that the closer we are to 100%, the more confidence we can have in the findings. The good surveyor determines how the nonrespondents might affect the observed results and establishes their characteristics. In Section V.B of Chapter 10 we give show how this was done in a higher education survey.

Governmental surveys often have a high response rate because they are subject to public oversight, have relatively generous resources, often are continuing surveys, and are conducted by professionals with a lifetime commitment to statistical analysis. For example, the U.S. Bureau of the Census' monthly sample survey of the U.S. population has a response rate near 95% [17, p. 197].

Unfortunately, some survey reports fail to even mention the response rate. The report of a professional polling organization discussing registered Democrats says only:

> A total of 850 registered Democrats in New York State were interviewed by telephone in late April. These respondents were randomly selected from current voter registration lists in 100 Election Districts. The 100 Election Districts were selected by a random method according to which an Election District's chance of being selected was equal to its proportion of the total number of combined votes cast in the 1976 U.S. Senate and 1978 Gubernatorial Primaries. [18]

This description, with its careful (and correct, for these purposes) method of choosing the sample, cannot be taken to mean that the actual sample obtained and used was valid. We do not know what proportion of the original sample is represented by those 850 respondents.

Aging in America: Trials and Triumphs [19] is based on a telephone survey of the U.S. population age 60 and over. The purpose of the study was ". . . to identify the negative forces in society that older Americans perceive hurt them the most—and the positive forces that they perceive provide their most effective support systems" [19, p. 2].

The first summary finding is that ". . . 67% of senior citizens say they always feel useful; 65% report strong self-images; 61% feel that, in general, things are worthwhile; 56% are serene; 52% show a high degree of optimism" [19, p. 13].

These numbers—67, 65, 61, 56, 52—imply accuracy and precision. We can evaluate them because this report gives some information on the responses. What do we find in regard to the responses? The survey reached 902 individuals age 60 and over; how many in the original sample design were not "reached" we do not know. Of these 902 contacts, only 514 supplied whole or partial data. Only 481 respondents supplied full data! Assuming that the original design called for a sample of 902, only 53% responded.

And what characteristics might the non- or partial-respondents have in common? Could they be the individuals with insufficient self-image to participate in a phone interview or to complete one? Could they be the physically disabled who feel useless? Could they be the mentally agitated who are not serene? Could they be the depressed who have neither optimism nor the capacity to deal with a stranger interviewing them about their feelings on the telephone?

Individuals over age 60 living in institutions were not contacted, nor were those without telephones. This occurred either from poor sample design or the effort to hold down costs. The bias is in the direction of a more positive result than if the whole universe were sampled and the nonrespondents evaluated. Older people who are living in institutions are more likely to be ill or disabled and hence less likely to be as optimistic as the survey report portrays older people. As we discussed earlier, poorer people are less likely to have telephones and we suspect they also are not as optimistic as the portrait given in *Aging in America*.

Furthermore, no comparisons were made with the population under 60 years of age. We need that information to make a useful interpretation of the survey results.

In the 1979–1980 survey of the membership of the Airline Passengers Association, Inc., all members of the association were sent questionnaires, but only 32% of the U.S. membership and 18% of the international membership responded [20]. Thus, 68% of the U.S. membership and 82% of the international membership did not respond to the survey.

Question 4 of the survey asked recipients to name the U.S. airline they would choose for travel within the United States. American Airlines led the list. Of the 11,931 respondents to the survey 10,227 replied to this question. Of those 10,227, 2446 chose American Airlines. Seventeen hundred and four respondents did not answer Question 4 or indicated no preference. They may or may not have preferred American. What can we say of these results? That *of those who responded* at least 20.5% (2446 divided by 11,931) explicitly choose American Airlines for travel within the United States.*

But what of the nonrespondents? If most of the nonrespondents preferred American Airlines, then this preference is seriously understated; if few of them preferred American Airlines, then this preference is overstated. The preference may lie somewhere between these extremes, but we have no way of knowing or even estimating the precision of the result as it would apply to all "frequent flyers."

The association gives no interpretation of the results in its report, and there is no basis for calling the survey a misuse. However, when American Airlines uses these results in its advertising, it is a subtle misuse, for their advertisement reads: "For the third straight time, American Airlines has been voted the number one choice for domestic travel in the Airline Passengers Association survey of the most demanding passengers in the sky: frequent flyers."

The advertisement is literally true, the votes received in the survey chose

---

*This value is enough to give American first place among the respondents. The proportion (among respondents!) might be even higher, depending on how many who stated "no preference" or did not respond would, in fact, choose American.

American Airlines as number one. But those votes were from only about one-third of the membership of the association. We have no idea how the membership as a whole feels. And neither we, nor the Airline Passengers Association, nor American Airlines, knows from this survey whether "the most demanding passengers in the sky" vote American Airlines number one. After all, these "votes" were not cast in an election, where we accept the choice of those voting (no matter how small a proportion of those eligible). This is a survey to determine the opinion of frequent flyers with regard to their choice of airlines, and because of the high proportion of nonrespondents (over 68%) we still don't know their opinion.

IV. More to Come

We close the first part of our discussion of surveys and polls here and will summarize both parts of our discussion in Chapters 9 and 10 at the end of Chapter 10.

10
Surveys and Polls: Analyzing the Data

The person who must have certitude, who cannot embrace conclusions tentatively, should not be engaged in social scientific research.

Norval D. Glenn

"Tut, tut, child," said the Duchess. "Everything's got a moral if only you can find it."

Lewis Carroll

I. Introduction

In Chapter 9, we looked at some of the ways in which the basic data in a survey can be so defective that the reported survey results are misuses of statistics. But suppose the data are not deficient, the questionnaire is not biased, the sampled and target populations agree, the counts are double- and triple-checked, and so forth. Is this enough? A mathematician would say that these conditions are *necessary*, but not *sufficient*. For sufficiency—that is, for the reported results to be sound and meaningful and thus serve your purposes, the subsequent analysis must be correct.

Fortunately, in many cases, even the statistically unsophisticated reader can evaluate the results and confirm or refute reported results. In this chapter, we discuss some—but not all—of the factors in the analysis of survey data. We conclude with an example showing how you can do your own evaluation.

The critical issue in many survey analyses where the data are obtained from a sample is the error resulting from the process of sampling itself. Although this involves some "mathematics," you should be able to "do it yourself."

II. Variability

A. What Is Variability?

By its very nature, a sample cannot give results that are guaranteed to be the same as the results of a census of all the individuals in the universe of interest. For example, try to determine something as simple as the proportion of males in some universe. The sample may give a proportion that is exactly the same as the universe (unlikely), near the value (likely under the right circumstances), or very far from the value (unlikely). This uncertainty—which is inherent in the random process of choosing a sample—we call *sample variability*. If you choose the sample of individuals to be queried in a suitable random manner, then you can estimate the precision of the sample results to measure the variability of those results.

Precision talks! *Precision* tells how well the measurement has been performed. Let's use the simple measurement of length as an example of precision. Then we will relate it to sampling variability. The meter is defined as 1,650,763.73 wavelengths of red-orange light given off by krypton-86 under certain specified conditions. If you mark this distance off on your meter stick to represent one meter, then your meter stick is highly *accurate*. If this meter stick has millimeter markings on it and you use it to measure length, any measurement will be to plus or minus one millimeter; this is the *precision* of your measurement. If it had only one centimenter markings, any measurement would be to plus-or-minus one centimeter, and the measurements would be less *precise*. If the meter stick has one millimeter markings and, thus, is precise to one millimeter, you cannot be sure of its *accuracy* unless you checked it as described above, or compared it with a reliable "standard" which was checked in that way.

In surveys, precision is a function of the sampling process, not the markings on a scale. Survey precision is usually measured in terms of a numerical interval higher-than or lower-than the reported result, analogous to the "plus-or-minus" one millimeter in our meter stick example. Thus, a survey result may be reported as "40%, plus-or-minus five percentage points." If the media has observed good practice, its reports will include this information, sometimes called the *margin of sampling error*.

1. *The Precision of a Survey Estimating a Proportion*

Depending on how a sample is chosen, even a nonstatistician may be able to determine the precision. In any case, an experienced statistician can make the computations. A good survey report gives any reader enough information to verify the reported precision. When this information is not provided, the surveyor prevents the reader—whether a statistician or not—from calculating the precision and prevents the reader from evaluating the work.

As mentioned before, when we get the precision of a sample survey we use a "plus-or-minus" concept, just as we do for measurements of distance, weight, and other physical quantities. Consider making a survey of a given universe to determine the proportion of males, and assume that the proportion in the universe is one-half, .50. If you took a great many samples by a scientifically determined sample, you would certainly get many different sample proportions, such as .40, .49, .51, .54, and so forth, because of chance variation.* Of course, you never do take a series of identifical samples—you take only one. How do you evaluate the precision of that one sample?

We take our one result and put a "plus-or-minus" interval around it. It is customary in survey sampling to choose this interval for a "95% confidence level." This means that, if we repeatedly took the same sample and made a plus-or-minus interval in the same way, 95% of the time our interval would contain the true value. Thus, if we interviewed 100 respondents and got a result of .40 for our sample of the universe for which the true value of the proportion of the surveyed characteristic was .50, we have an interval of .30 to .50.

Of course this is a hypothetical situation, since if we knew the true value we would not need to make a survey. The value of considering this hypothetical situation is that it is the basis for our evaluation of the process of surveying a universe *once*, and evaluating the precision of the result of that survey.

How did we arrive at the conclusion that the interval is from .30 to .50? Let us show you how the interval was calculated. The unit of measurement is the *standard error*, which depends primarily on the sample size. In the case of samples taken in surveys of the type we have been describing, where the sample size is a small fraction of the universe size, the proportion of the universe taken in the sample is unimportant.† What *is* important is the sample size. Consider those simple cases which are most common in popular survey-making. If you make plus-or-minus two standard error intervals around the survey result, then you will have a 95% confidence level.

---

*You can make your own experiment in which you make one toss of 20 coins (or more if you have the patience) and count the number of heads. Since the probability of getting a head is one-half, this is analogous to taking one sample from a universe in which the proportion of heads is one-half. Repeat the toss several times. Each time you make a toss you should get a different proportion of heads. Note the mean and spread of the proportions. From this you can get an idea of the precision of the sampling process. Try it again with 40 coins, and note once again the mean and spread of the proportions. This is the best way to get an appreciation of the nature of confidence intervals and precision in sample surveys.

† When the sample size is a "large" fraction of the universe size, the standard error described by us is multiplied by the *finite correction factor*, which is less than one. How large is "large"? If the sample size is one-hundredth of the universe size, then the finite correction factor is .995. If the sample size is one-tenth of the universe size then the finite correction factor is .95; when the sample size is one-half of the universe size then the finite correction factor is .70. How often do you see surveys in which the sample size is more than one-tenth of the universe size?

How do you calculate the standard error? For most simple surveys involving proportions, it is remarkably uncomplicated. The standard error is

$$\sqrt{\frac{(\text{Observed proportion}) \times [1 - (\text{Observed Proportion})]}{(\text{Sample Size})}}$$

Thus, if you did not know the true proportion of males in a universe, took a random sample of 100 persons from a universe to determine the proportion of males, and got a sample proportion of .40, the standard error would be:

$$\sqrt{\frac{(.40)(.60)}{100}} = \sqrt{.0024} = 0.05$$

The width of the 95% confidence interval is:

$$\pm 2 \times .05 = \pm .10$$

Thus the 95% confidence interval is .40 plus-or-minus .10, or .30 to .50. This precision is the consequence of the small sample size. The surveyor must decide whether this precision is satisfactory.

2. *The Precision of a Survey Estimating a Mean*

If the result is an average (the arithmetic mean) and not a proportion, the computation is similar. The 95% confidence interval is plus-or-minus two standard errors on either side of the sample mean. But you need to know more than the sample size and the observed sample mean. You must also know, or be able to estimate, the standard deviation of the observed variable in the universe. If the specific data values obtained in the survey are available, you can estimate the standard deviation by a computation that can be made on a pocket calculator. You can then get your estimate of the standard error by dividing the standard deviation by the square root of the sample size. A good survey report will give either the standard deviation or the standard error if the original data are not given.

To measure sampling variability in this way, the sample must be a "random sample" where every member of the universe is given an equal chance of being chosen.* If the sample was self-chosen, haphazard, or directed by the judgment of the surveyor, you cannot say that the 95% confidence interval is the observed value plus-or-minus two standard errors. In complex samples, such as some of those used by the U.S. Bureau of the Census and the U.S. Bureau of

*This is a simple definition which is adequate for the purposes of our discussion.

Labor Statistics, variability can be measured but the procedures are more complicated than the simple ones we have described.

Of course, you may not have to calculate the precision. Many survey reports include an estimate of precision calculated by the author or a professional statistician, as is done by some media. The following excerpt from an Associated Press release, which was quoted in full by the *Stamford Advocate*, is a model of good media reporting on surveys:

> Respondents in the Media General–Associated Press poll included a random scientific sampling of 1,412 adults across the country Sept. 1–7. As with all sample surveys, the results of the Media General–AP telephone poll can vary from the opinions of all Americans because of chance variation in the sample.
>
> For a poll based on about 1,400 interviews, the results are subject to an error margin [confidence interval width] of 3 percentage points either way because of chance variations in the sample. That is, if one could have questioned all Americans with telephones, there is only 1 chance in 20 [95% confidence] that the findings would vary from the results of polls such as this one by more than 3 percentage points. [1]

B. The Use of Variability to Interpret Survey Reports

1. *Family Income*

Good surveyors evaluate precision or give their readers the means to make their own evaluation. For example, the U.S. Bureau of the Census published data from its March 1983 sample survey (the CPS) which included both the mean and median incomes* and measures of variability [2, Tables 1 and 4]. For white households, the mean income in 1983 was $26,455 and the standard error, $102. Hence the 95% confidence interval for the mean income of white households is from $26,251 to $26,659.†

Table 10.1 shows the results of this survey for the three major racial groups. As you can see from this table, none of the confidence intervals overlap. Thus, you can reasonably conclude that white household mean income is greater than Hispanic, which in turn is greater than the black household mean income. If the sampling variability had been so great that the intervals overlapped, even though the sample means were different your conclusion would be different. In that case you would have to conclude that the survey gives you no basis for

*As discussed in Chapter 7, the median income is the "middle value," that value which splits the data into two equal parts. For all practical purposes, half the families have less than this value, half have more than this value. Variability can be calculated for the median as well as the mean, but we have not discussed how this is done.

†If this survey were repeated an infinite number of times (an obvious impossibility), the interval calculated this way would contain the true mean of all white households 95% of the time.

Table 10.1 Mean Family Income, Standard Error, and 95% Confidence Limits

Race	Mean income	Standard error	Lower 95% limit	Upper 95% limit
White	$26,455	$102	$26,251	$26,659
Black	16,531	202	16,127	16,935
Hispanic	19,369	363	18,643	20,095

saying that there are differences in mean household income among these three groups.

To say that "in 1983 the income of the average white household is $26,455" is an incorrect statement—a misuse of statistics. The correct statement is that, based on our survey, we are "95% confident that the mean income of *all* white households is between $26,251 and $26,659." And since stating these figures to the single dollar implies a precision that doesn't exist (called *spurious precision*) it is better to say "between about $26,000 and $27,000," which also is easier to grasp.

3. *Unemployment*

A *New York Times* article reported that "some states scored heartening declines in joblessness; New York's unemployment rate was 6.8% compared with 7.4% in October" [3]. The "heartening decline" is .6%, less than half of the variability of 1.5%, which is two standard errors [4, Report 571, Table A-8, p. 97]. The decline of .6% is meaningless in light of the sampling variability, not "heartening."

Headline writers for U.S. government publications are not necessarily more careful than headline writers for newspapers. The U.S. Bureau of Labor Statistics issued a report (May 6, 1983) with the headline, "Unemployment Rate for New York City Declines to 9.2% in April." The reported decline was 1.8%, but the 95% confidence interval width for the monthly changes at this time was more than plus or minus 2.5%. The change between March and April may well be only a chance effect due to sampling.

4. *Is It Always 95%?*

So far we have always made confidence intervals with a 95% confidence. This is the customary level of confidence used to define the confidence intervals in surveys reported by newspapers such as the *New York Times*. You are not obligated to use this level; if you can live with a lower confidence level, then you will have a narrower interval. For example, the 68% confidence interval for the mean white household income in 1983 is from $26,353 to $26,557, since the width of the 68% confidence interval is plus-or-minus one standard error.

Do you require even more confidence than you get from a 95% confidence interval? Then you can use a 99.9% confidence interval; mean white household income is from $26,149 to $26,761, since the width of the 99.9% confidence interval is plus-or-minus three standard errors. The narrower the range, the lower the confidence level, and the lower the reader's confidence in the reported numbers.

III. Listen to the Professional Pollsters

Professional pollsters have an excellent reputation for describing sources of errors in their statistics which could lead to misuses. They often are leaders in conducting tests to reduce the number of such errors and innovative in finding new approaches. While the totally error-free survey may always elude us, professional pollsters often succeed in reducing errors to tolerable levels and giving the reader sufficient information to evaluate the results.

Professional pollsters have publicly given much good advice, which, if taken by their clients, the public, and others, would greatly reduce misuses of statistics. For example, on nonsampling errors (bias, those involving accuracy) the U.S. Bureau of Labor Statistics tells us:

> Nonsampling errors in surveys can be attributed to many sources, e.g., inability to obtain information about all cases in the sample, definitional difficulties, differences in the interpretation of questions, inability or unwillingness of respondents to provide correct information, inability to recall information, errors made in collection such as in recording or coding the data, errors made in processing the data, errors made in estimating values for missing data, and failure to represent all sample households and all persons within sample households. [5, p. 198]

Daniel Yankelovich, formerly president of Yankelovich, Skelly, and White, Inc., a public-opinion research organization, warns readers about an aspect of political polls we have not discussed:

> As opinion polls have grown more influential in recent years, the public needs to gain a better understanding of a seeming paradox: The polls are almost always accurate in the narrow sense of reporting what cross-sections of Americans actually say in response to particular questions at a given time. Unfortunately, though, even accurate polls can be misleading because what the people say is often not what they really mean.
>
> There is nothing mysterious or unsatisfactory about this. It is not a technical problem having to do with sampling, the phrasing of questions, of the tabulation of statistics. Nor is it a moral problem. People almost never lie outright in polls and they virtually never seek to mislead. When faced with the eventuality of an important decision, most people do not sort out their convictions until they have spent weeks or months 'working through' their feelings and attitudes. . . . at any point along the way a public-opinion poll may catch an attitudinal 'snapshot' of the public in the act of making up its mind. [6]

During the 1980 presidential election, the League of Women Voters scheduled a pre-election debate between candidates Carter and Reagan. The League could not decide whether John Anderson, a third candidate, should also be invited to the debate. The League decided that it would invite John Anderson only if "the polls" showed that at least 15% of respondents supported him. What the League has done here is to set up a *decision rule*, a quantitative basis for making the decision as to whether or not to allow John Anderson to participate in the debates. There is nothing wrong with the League making and using a decision rule based on some criterion of public support.

However, they chose to use a rule based on "the polls," because they were seeking a criterion capable of objective evaluation to measure the extent of Anderson's public support. Were "the polls" an objective measure?

Referring to this issue, an opinion analyst and the editor of the book *Polling on the Issues,* Albert Cantril, saw many flaws in this precedent-setting use of the polls as a part of the political process itself:

> What makes pollsters uneasy are two fundamental misunderstandings about public-opinion research: that it should play a formal role in the matter of whether Mr. Anderson is invited and that it can provide the finely calibrated measures that the League expects. . . . Most pollsters share the conviction that when it comes to the conduct of the public's business, there are two appropriate vehicles: the ballot box and the representative institutions of government. . . .
>
> The results of a poll are nothing more than the pattern of answers people give to the questions asked of them. The findings of a poll may be reliable in a statistical sense, but have nothing to do with public opinion if the questions have been poorly phrased. . . . Because of the initiative that resides with whoever frames the questions, pollsters can play a central role in defining political issues.
>
> Consider Louis Harris's question that asked people to "suppose Anderson had a real chance of winning" before soliciting their preference among Ronald Reagan, President Carter and the Congressman [Anderson]. The results produced a figure for Mr. Anderson that was 11 percentage points higher than in the standard three-way "trial heat" question. [7]

We feel that the officials of the League were abdicating their responsibility in the decision-making process as well as setting a dangerous precedent. This issue was too important to settle by one throw of the pollster's dice. They would have served the political process better if they had made a group decision based on a study of all the available information from the several pollsters and on their own knowledge of the political process.

Professional pollsters can give guidance in designing a survey. We don't expect most of our readers to design and carry out a survey, but the knowledge of good practice from the viewpoint of a professional can help in two ways: (1) by increasing the knowledge base for evaluation of surveys; and (2) if you wish to use a survey to get information, in making an informed judgment of the proposals and plans of those who will carry out the work. *Development of Survey Methods to Assess Survey Practices* is a publication of the American Statistical Association which succinctly gives a professional's viewpoint of surveys of human populations.* The findings and discussions are relevant to all types of surveys.

Among the observations reported from a pilot study of surveys are:

> 1. Fifteen of 26 Federal surveys did not meet their objectives. . . . technical flaws included high nonresponse rates, failure to compute variances [basic to determining standard errors] or computation of variances in the wrong way, inclusion of inferences in the final report that could not have been substantiated by the survey results, no verification of interviewing, and no data cleaning.
> 2. Seven of the ten non-Federal surveys did not meet their objectives.
> 3. Samples were, for the most part, poorly designed.
> 4. Survey response rates were difficult to collect and compare.
> 5. Quality control over data-processing operations varied considerably. . . .
> 7. One would hope that survey results bearing on controversial subjects or public policy would include enough evidence to support reported findings. . . . This is not generally the case. Causal inferences were often made in cases where variables were correlated in such a way that causal relationships could not have been determined by a survey of the type used. [8, p. 13]

Fortunately, this publication includes recommendations on how to conduct a survey as well as criticisms based on its studies of existing surveys. We encourage you to bear the above observations in mind and to think critically about surveys and their results.

IV. Evaluating a Survey—How You Can Do It

To show you how to evaluate a survey, we analyze a New York Times/CBS poll of voters. Fortunately, the *New York Times* now publishes some detailed information on how it conducts its polls.

We now quote the relevant section of the article, make our observations, and ask questions which help in judging the results:

> The latest New York Times/CBS New Poll is based on telephone interviews conducted Aug. 5 through 9 with 1,616 adults around the United States, excluding Alaska and Hawaii, of whom 1,188 said they were registered to vote. [9]

The sample was chosen by the use of a computer, which first selected a sample of all of the telephone exchanges in the United States, choosing exchanges so that each region of the country was represented in proportion to its population.

---

*Available from the American Statistical Association, 806 Fifteenth Street, N.W., Washington, DC 20005.

For each chosen exchange, telephone numbers were formed by random digits, thereby giving access to both listed and unlisted residential numbers.

Is the sample representative of the universe in those relevant characteristics for which we have information from other sources? Of the 1616 adults contacted, 1188 said they were registered to vote. This proportion is 74%, close to the value of 70% reported by the U.S. Bureau of the Census [10, Table 438]. We list some observations and questions concerning the survey.

1. Will the results be affected by the omission of homes without telephones? As we have shown, poorer residents and minority members have disproportionately fewer telephones in the family home [11], but does it matter in this survey? We lack definitive general information on how telephone interviews differ from personal interviews, but the researcher should estimate the effect of omitting poorer individuals in any given case. What would they respond to this question? No information is given to help answer this question.
2. How many of those who received phone calls refused to answer? No information is given.
3. What was done if a business phone was reached by random dialing? No information is given.
4. What was done if a juvenile answered the phone in a private residence? No information is given.
5. What was done if no one was home when the call was made? No information is given.
6. If the home which was reached contained two or more adults, who was interviewed for the survey? No information is given.
7. Does the sample of telephone exchanges and the method of choosing numbers appear to give a good random sample of homes having telephones? We would have to examine the sampling design for exchanges in detail before drawing any final conclusions.
8. Why did the surveyors weight the results, and not publish the original data? The results have been weighted to take account of household size and to adjust for variations in the sample relating to region, race, sex, age, and education. If the random number selection and the sample of telephone exchanges gave a proper random sample, why was it decided to weight the results? What were the weights and how were they determined? No information is given.

Without answers to the unanswered questions raised above (and surely more which would be apparent on seeing the complete report), we cannot fully evaluate the results.

V. Summary of Chapters 9 and 10

A. Common Pitfalls and Cautionary Recommendations

A great part of the statistics which the public sees and which affects public policy is based on surveys. Someone formulates questions, decides on a sample, and asks those questions of people in the sample. There is much room for misuse of statistics in this process.

Below are six major pitfalls and suggestions for the careful reader who wishes some guidance:

1. *Leading questions—those which consciously or unconsciously are designed to obtain desired answers*. The way to deal with this pitfall is to find and evaluate the original questions. Much good statistical survey work is being done and the major professional pollsters and many publications make a point of giving the exact wording of the questions. If the result of the survey is important to you, it is worthwhile to seek out the questions, even if some extra effort is required.
2. *Surveys which are really requests for contributions and attempts to gain "scientific" support for one viewpoint*. These surveys usually have leading questions linked to requests for contributions. Such so-called surveys are not surveys and cannot, and should not, be given the credibility that you give to a professional survey or one in which the surveyor has tried (even if only partially successful) to be unbiased. However, there is information in the results of such a misuse of statistics, for the nature of the responses tells you something about a particular group of respondents.
3. *Self-selected samples*. Self-selected samples tell you only about the characteristics of those who have selected themselves. In general, the results cannot be projected to any other universe.
4. *Nonrespondents*. Know what the response rate is. Determine how the nonrespondents might affect the observed results. Good surveyors establish the characteristics of the nonrespondents to estimate what effect they might have. Look for such efforts in the report. In the example in the following section, we show how to deal with nonrespondents.
5. *Failure to take into account the effects of variability*. If the report does not give a measure or estimate of variability, then compute one as best as you can. If the sample size or other information needed to compute variability (for example, the standard deviation) is not available and the survey results are important to you, ask for them.

6. *Misinterpretation of the results, either in the report or in subsequent media reporting.* You can avoid this pitfall by reading the full body of the report and checking for logical consistency.

B. Are We Asking Too Much?

In a real situation, no survey will be perfect. Nor do surveys need to be perfect; they only need be good enough to serve the purpose for which they are undertaken. Imprecision and inaccuracy may be tolerable and it is important to know "how good is good enough."

One of the most frequent sources of inadequacy in surveys is, as we have mentioned, a high rate of nonresponse. If a survey has a high proportion of nonrespondents, there is always the possibility that the results might have been significantly different if the nonrespondents had responded. Below we describe a case in which the surveyors struggled, and succeeded, in overcoming this problem, providing a useful model for other surveyors.

In a study described in *Negro Higher Education in the 1960s* [12], the survey universe was all of the Negro high schools in states where less than 8% of blacks were attending high school with whites in 1964–1965. The first mailing of questionnaires went to 1831 high schools. After this mailing the surveyors discovered that some schools had closed, others had changed their grade levels, and some on the original list were white schools. The returns from the first mailing were so low that a second mailing was made. Only 203 high schools responded to the first two mailings and:

> Given the low initial rate of response, it was beyond our financial means and time limitations to secure returns from the two thirds or better of all the schools necessary to assure a reasonably representative sample.
>
> Instead, we attempted to obtain a 100 per cent response from a sample of fifty of the schools on the list. [These responses] would provide an index of the representativeness of the larger sample when compared to . . . the larger group of returns. . . .
>
> After extensive follow-up mailings, plus numerous phone calls, we managed to account for all fifty high schools. For forty of them we obtained useful questionnaire returns, five were White high schools, and five were no longer high schools. [12, pp. 204–5]

These results were then systematically compared with the results from the larger sample with its low response rate, and the final data were adjusted to account for the information gained from the sample of the nonrespondents. All of this occurred under conditions of limited financial means and time. If the will to make sure that the survey is not deficient is there, the means and the time can be found.*

*For a more complex example in a large-scale survey, see Reference 13.

11
The Law of Parsimony: Ockham's Razor

Out of the clutter find simplicity.
Out of discord make harmony.
Out of difficulty find opportunity.

Albert Einstein

You go on the tennis court to play tennis, not to see if the lines are straight.

Robert Frost

I. Introduction

William of Ockham (1280–1349) was an English scholastic philosopher. In an era when the most convoluted arguments were used to prove how many angels could dance on the head of a pin, he proposed his principle of parsimony [also known as "Ockham's razor"], which said: "What can be done with fewer (assumptions) is done in vain with more." Unfortunately, modern statistics is often in need of a shave.

Translating Ockham's razor for the world of statistical analysis, this becomes: the simplest procedures which can be used to solve a problem are the preferred ones. Those who obfuscate solutions by using complex procedures when simpler ones will work just as well are guilty of major misuses of statistics. Such misuses may be satisfying to the soul of the user and bolster the ego (See how smart I am!). Or they may obscure the analysis (I don't want anyone to see how shaky my work is). Sometimes, these misuses even lead to a job promotion for the perpetrator (See how smart I am, boss!):

> Complexity and obscurity have professional value—they are the academic equivalent of apprenticeship rules in the building trades. They exclude the outsiders, keep down the competition, preserve the image of a privileged or priestly class. The man who makes things clear is a scab. . . .
>
> Additionally and especially in the social sciences, much unclear writing is based on unclear or incomplete thought. It is possible with safety to be technically obscure about something you haven't thought out. It is impossible to be wholly clear on something you do not understand. Clarity thus exposes flaws in the thought. The person who undertakes to make difficult matters clear is infringing on the sovereign right of numerous economists, sociologists and political scientists to make bad writing the disguise for sloppy, imprecise or incomplete thought. [1]

A wave of unnecessary complexity can drown the audience with statistical misuse.

In an amusing satire, Edi Karni and Barbara K. Shapiro describe what can happen when statisticians disregard the principle of Ockham's razor:

> [The investigation of the Committee on the Mistreatment of Raw Data—COMRAD] . . . unmasked such nefarious schemes as employment of third-degree autoregression processes and, in what may be the report's most revolting disclosure, the brutal imposition of third-degree polynomial structures. Maximalist methods also include the use of first and even second differencing, which according to eye witnesses, often reduced the data to a totally unrecognizable state. [2]

II. Examples

A. A Lack of Sufficient Statistical Discrimination

Naive use of statistics (and attendant complexity) is illustrated by the report of the Ad Hoc Committee to Implement the 1972 Resolution on Fair Employment Practices in Employment of Women of the American Anthropological Association [3]. The purpose of the committee was to ascertain whether some academic anthropology departments discriminated against women in their hiring practices. The known facts were: (1) for the years 1973–1977, the proportion of all doctorate degrees in anthropology that were awarded to women; (2) the proportion of all hirings that were women, school by school, in departments of anthropology; and (3) the proportion of women faculty members in each school that were women.

A nonstatistician (or the owner of a suitable razor) would simply have compared the proportion of women in all hirings, school by school, with the proportion of doctorate degrees awarded to women overall. The investigator would then know which schools hired a smaller proportion of women than the national average. Then you can search for the reasons why these schools have a lower proportion to determine if discrimination has taken place. A neat, close shave by Ockham's razor.

Instead, the committee devised its own measure. Someone decided to cross-multiply the two percentages: the proportion of women in the faculty and the proportion of newly-hired faculty members who were women. This is comparing oranges with oranges, for the numbers must overlap. The resulting statistical tables were most imposing and may have misled people. We hope that some people were able to look behind the complex presentations and draw reasonable conclusions. Naive statistics and unnecessary complexity can only have made the task more difficult [4].

B. In the Digs

No field is immune. Even archaeologists have disregarded Ockham's principle. Writing of this, David H. Thomas, of the Anthropology Department of the American Museum of Natural History, describes how archaeologists have used—and overused—complex statistical methodology:

> Too much of the methodological "new archaeology" seems involved with adapting show gimmicks, which function as symbols to attract adherents and amass power . . . simulation studies, computer methods of typology, and the numerous "models" that blight the archaeological literature of late—random walks, black boxes, Markov chains, matrix analysis, and so forth. . . . Quantitative techniques can (and do) provide valuable tools when properly used. Specialized quantitative methods become bandwagons . . . when they become ends in themselves, playthings that serve as roadblocks to real understanding. [5, p. 235]

As an example of how the law of parsimony was violated, Thomas describes Joel Gunn's 1975 analysis of C. Melvin Aikens's original findings, published in 1970, discussing an excavation at Hogup Cave in Utah [6]. Gunn stated that the purpose of his re-analysis of Aikens's data was to determine "to what extent environment causes cultural change" and Gunn is quoted by Thomas as saying "It was considered important to keep the substantive aspects simple and to concentrate on methodology." This is our first warning signal!

Thomas describes how Gunn "transformed (Aikens's data) by $\log_{10}$ to reduce skewing effects of very high artifact counts, standardized to a mean of zero and a standard deviation of one, then similarity coefficients were computed and factored." Thomas goes on to note that the statistical relationship which emerged "was patently obvious to all who had read Aikens's analysis." Then why was it processed and massaged to such great lengths when the statistical conclusions were not altered?

Next, Gunn carried out factor analyses and multiple regressions on the data. Thomas says, "The present analysis is so far removed from archaeological reality that the numbers take on a life of their own. And the conclusions offer us nothing new" [5, p. 240]. Indeed, Thomas was truly amazed at one of the major conclusions, which stated that "habitat causes a cultural change of 50%." What on earth does this mean? The archaeological specifics that Aikens had considered had been transformed, by statistical maneuvering, into archaeological numbers, largely without meaning. Gunn's analytic overkill illustrates a too-frequent academic misuse: Concentrate on the methodology, the substance be damned.

C. Analyses of Nonexistent Data

1. Death Rates

Statistical analysis of death rates is another field where Ockham's razor could be used. Here, it is often the lack of good original data which seduces statisticians into overusing the data which are available.

If death rates can be calculated, then it is possible to calculate a life table and measure the average number of years which the newborn can be expected to live.* But in many parts of the world it is not possible to get the basic data necessary for the calculation of death rates. If the basic data—the number of people in each age category and the number of deaths that occur in each category—are unknown, then we cannot calculate death rates. What is to be done?

The demographer W. Brass devised an analytical method to develop the probability of death which involved asking women about the numbers of their dead and living children [7–9]. The proportion of dead children (of all born), when combined with knowledge of the age of the mother and the ages of the children, gives the death rate. The life expectancy at birth can then be estimated, using a series of complicated procedures. This description is an oversimplification, but it does describe the general methodological idea.

There are many assumptions that must be made to support these complex analytical computations. Michel Garenne gives a list of eight such assumptions [10]. We feel that the basic problem with this methodology is that the respondents must give *correct* information about their ages, the number of children born to them, and the number who died, according to age at death. We do not go into the problems with the other assumptions here.

Michel Garenne proposed a closer shave in his article. After carrying through all the necessary Brass-method computations, Garenne concludes:

> In this example from Tropical Africa, it seems that at this level of knowledge, data on proportion of children dead do not permit the computation of reliable estimates of infant and child mortality. . . . Inaccuracy of data on age as well as underestimation of children deaths will give no more than a rough idea of the level of mortality . . . *Why not stay with the proportions of children dead? At least by themselves, they give an idea of the level of mortality* [italics ours]. . . . As long as there are no accurate data in Tropical Africa, one cannot expect to have more than ''ideas'' of levels and differentials in mortality. [10, p. 129]

---

*See the Appendix.

We are not casting doubt on the Brass methodology in general. The Brass methodology need not be shaved if a country has adequate data for the application of the Brass methodology and lacks the necessary data for computing death rates.

2. *Answers Without Data*

Wassily Leontief, the winner of the 1973 Nobel Prize in Economics, is troubled by the tendency of economists to build models without using actual data. He analyzed the articles which appeared in the *American Economic Review* between March 1972 and December 1981 and found that over half were:

> . . . mathematical models without any data. Year after year economic theorists continue to produce scores of mathematical models . . . without being able to advance, in any perceptible way, a systematic understanding of the structure and the operations of a real economic system. [11]

We agree—models that are not tested with data constitute misuses. Leontief further comments:

> Thirty-six percent [of research papers studied] also contain attempts at empirical implementation of these intricate theoretical constructs. Such attempts involve routine application of elaborate methods of indirect statistical inference applied to a small number of . . . indices. . . . Only two of the 44 researchers saw fit to engage in the grubby task of ascertaining by direct observation how business actually arrives at [its assessments of government behavior and consequent effects]. [11]

The "lack of substance and precision in much current economic literature" bothered Richard Staley, another observer:

> I once described this phenomenon to some graduate students as a procedure of piling estimate on top of conjecture, declaring the whole to be an axiom based on the author's reputation and then using this "base" to launch still further estimates and pseudo-precise conjectures. . . . Personally, I tend to opt for the philosophy that less may be better under these circumstances. [12, p. 1204]

3. *Hybrids*

Sometimes overcomplexity is mixed with oversimplification to produce a hybrid misuse. Gerald A. Bodner was involved in a court case involving the difference in salary between a group of male faculty members and a group of female faculty members [13]. The female faculty members claimed that they were receiving lower pay even though their qualifications were just as good. Were the salary differentials due to intentional sex discrimination? The plaintiffs' study produced regression equations which included as many as 98 independent variables. Do enough data exist to support that many variables? Any formula which attempts to cover that many variables indicates the need for a closer shave with Ockham's razor.

On the other hand, the court approved an analysis which included oversimplification. Information which should have been included in the salary analysis was omitted. Bodner writes:

> Pre-hire years of teaching experience and pre-hire publications, both clearly relevant and important factors, were left out because, as the statistician for the plaintiff suggests, with the court's acquiescence, they are "adequately accounted for by the variables of age, degrees, and years between degrees." The accuracy with which data on age and degrees measure pre-hire publications and years of teaching experience may be a good indication of the accuracy of the entire statistical case. [13]

The principle of Ockham's razor does not imply that relevant facts can be omitted. Proxy or substitute information instead of actual data, however, can constitute an oversimplification, especially when the actual data can be obtained—as is the case with pre-hire publications and teaching experience. We aim for the simplest approach, but not at the cost of disregarding relevant information.

D. Less Is More

Wesley A. Fisher tried to find out whether marriage between persons of different ethnic groups in the Soviet Union was due to ethnic consciousness or to social and demographic factors [14]. He analyzed 14 different ethnic groups in the Soviet Union and defined eight characteristics with which he hoped to explain the amount of intermarriage between persons of different ethnic origins. He analyzed the data using multiple linear regression to obtain an equation with which you can predict the amount of intermarriage (as we discussed in Chapter 7). He judged his results to be definitive since, by using all eight characteristics as independent variables in the regression equation, he accounted for 74% of the variance.*

Reviewing these results, Brian D. Silver of Michigan State University showed that five of the eight independent variables included in the regression equation added nothing to the final explanation [15]. One serious difficulty with Fisher's original analysis is the use of only 14 cases for which data were available (that's how many ethnic groups he had) to support an equation with eight independent variables. It is well-known that, as the number of cases and the number of independent variables approach each other, the reliability of prediction by the independent variables approaches zero. To give an idea of how close Fisher is to the practical limit with eight independent variables and 14 cases, note

*This concept is discussed in detail in Chapter 7. You may recall that the amount of variance explained by the independent ("explanatory" or "predictor") variables in a regression equation is the coefficient of determination, R^2.

that if the number of cases and independent variables is the same, then the proportion of variance explained will always be 100%.

Silver's recalculations from the original data agreed with Fisher's results. He then rejected the five independent variables that were not significant in explaining the variance and computed the resulting regression equation based on only three independent variables. This equation, shaved down from eight to three independent variables, explained about 79% of the variance, compared to Fisher's 74%.*

Silver interpreted his results and concluded that, because two of the three independent variables which indisputably contributed to the explanation of intermarriage were religion (even in the Soviet Union!) and native language, ethnic consciousness and not social and demographic factors, were most important. This directly contradicts Fisher's findings. Indeed, religion alone explains 55% of the variance in intermarriage; people in the Soviet Union tend to marry coreligionists. Are you surprised?

E. More Can Be Too Much

In the previous example, explaining 70–80% of the variance was considered a significant indication of the ability of independent variables to predict. "Would you believe 99.9969% explained" is the title of a provocative paper by two Dow Chemical Company researchers who used a complex computer program to get a regression equation which they hoped would lead to a solution of a quality problem [16].

Their odyssey in Ockham's never-never land began with their involvement in a production quality problem. A customer found that a chemical product had an unpredictable service life and questioned the quality of the batches. Representatives of the company's relevant operations (production, engineering, research, quality control) assembled and agreed to collect all the available data "and submit it to the computer for analysis." Additional data were also collected. Sixteen independent variables were defined; the dependent variable was service life of the product. Data were collected for 22 batches. Additional "independent variables" were created by performing transformations on the original 16: the reciprocal of each original variable, the square, and the reciprocal of the square.

---

*(This footnote is for readers with a modest familiarity with multiple regression.) You may be surprised to see that a reduction in the number of independent variables increases the explained variance. Usually, adding another independent variable increases the proportion of explained variance, but the increase can be insignificantly small. However, when the number of cases is small and close to the number of independent variables, the "adjusted" coefficient of determination must be used, and it is possible for a reduction in the number of independent variables to increase the adjusted coefficient of determination. Less can be more!

The computer program for analysis of these data is a "stepwise" multiple regression program. To explain the variance, this program moves in "steps," adding independent variables to the linear multiple regression equation one at a time.* At each step, the program adds to the equation that one variable from the remaining set of independent variables which explains the greatest amount of the (as yet) unexplained variance. The program also drops previously entered variables from the equation when the inclusion of a new variable makes them less important. In the Dow case, this complex analysis resulted in the following finding: about 75–80% of the variance in service life was explained by independent variables [16].

But the researchers were suspicious of the whole complex process. So they drew numbers out of a hat at random to get values for all 16 independent variables for the 22 batches. When they submitted these random data to the same computer program, they found that 99.9969% of the variance in service life was explained by random numbers! This is significantly higher than the 75–80% which aroused their suspicions in the first place, and threw the whole process into doubt, for "Of course, with results like that out of a hat, it was no longer reasonable to make extensive plant revisions" [16, p. 45].

The authors offered some sensible guidance:

> Pressures to use the latest mathematical tools in conjunction with complex computer programs are often great. Response to this pressure is evident in the number of training programs available to people engaged in quality control work.
>
> Such people, however, are in a very real dilemma: They have need to learn and to use these modern techniques, while at the same time, they have the responsibility to see that these techniques are not abused.
>
> This responsibility is often extremely difficult to meet in a practical way—that is, without resort to statistical jargon. [16, p. 46]

F. Or Less Is Certainly Good Enough

In the 1960s, one of us worked on the design of an extensive weather forecasting system for military applications. The system went into the field as a complete laboratory contained in a standard trailer truck. It had sensing devices and gave forecasts based on complex analysis of a wide range of observed atmospheric phenomena. Over ten years later, one of our students reported on his experiences in Vietnam:

> I would like to interject [into a report on statistical forecasting methods] a short story which demonstrates what is, possibly, the true strength of the naive model.†

*We discussed the use of regression to explain variance in Chapter 7, Section A, Part 3.

†The *naive model* means that you forecast the next period value with the current period value. For example, if a restaurant using the naive model wants to forecast the number of customers to be served for the next day, it uses the number served the previous day as the forecast value.

> When I said I did not call the naive model by its name, it is not that I just happened on the process by accident and didn't even think that such a process existed. My term for the naive model was persistance [sic] and it was an important part of my life for a year. The fact is, we are assailed by the persistant model every time we listen to a weather FORECAST. Let me briefly relate the story.
>
> From January 1971 to January 1972, while serving in the United States Air Force, I was assigned to the 1st Weather Group, Tan Son Nhut Vietnam. 1st Weather Group had responsibility for producing the weather forecast for Southeast Asia and providing that information to all commanders in support of war operations. As chief of weather equipment maintenance I used to sit on all the forecast briefings. Two briefings were given. The first briefing was the persistance briefing, it rained yesterday over DaNang, it will therefore rain today over DaNang. The second part of the briefing was forecaster analysis. This took into account all of the information which was obtained from what, I assure you, was the most advanced weather equipment known to science. [This was the equipment described earlier.] *The weather forecast improved upon persistance by 0.02 percentage points* [our italics]. Both persistance and the forecaster analysis were correct to a significant level.

Weather forecasting has come a long way since the early 1970s. But at that time, for the purpose intended in that place, would saying more have been less?

G. Doing Little in a Big Way

"PIMS" (Profit Impact of Market Strategy) is a system designed to provide a factual basis for the business planning efforts of companies which participate by supplying data to a large data base. The initial work was done in General Electric's planning department in the 1960s using General Electric business data. In the early 1970s, developmental work on the system and data base was carried out in the Marketing Science Institute, which is associated with Harvard's Graduate School of Business Administration. Today the system's home is the Strategic Planning Institute, a nonprofit corporation governed by its member companies.

An executive of a large American corporation, one of the "Fortune 500," who does not want to be identified, supplied us with an internal report on his experience with PIMS, part of which exemplifies the need for Ockham's razor. While this report may seem highly critical of PIMS as a whole, it should be recognized that it is really one particular application and the work of a few particular individuals which is criticized, for as the authors of the report note, "PIMS represents a courageous venture into mostly unexplored territory."*

The authors found many flaws, which they called "statistical fallacies." We excerpt from their report concerning the "most serious single defect":

---

*The reader should also note that this report was prepared before 1977.

> The fundamental PIMS model is . . . a regression, in which return on investment [the dependent variable] is ''explained'' by about 70 ''independent variables,'' consisting of quantities obtained from the questionnaire, their ratios, and several expressions (some very complex) built from these ratios. [It] claims to explain 78% of the ''variance'' in return on investment from one business to another.

When results concerning the company's capital structure as it applied to the several operating divisions were received, the corporate staff spent several months

> . . . together with the appropriate business analysts, attempting to understand the structure of the model and to trace the path from our input data to the final results. We found that [several terms in the model] played an overwhelming role, both in the general structure of the model and in its application to our businesses. The difficulty was that the [results] did not agree well with the qualitative judgments of [the] analysts.

What the company's analysts discovered was that data of many kinds were input to the complex computer analysis system and were used to produce a regression equation which produced a value for the ratio of income to investment. But one of the company's analysts observed

> that, as a matter of high school algebra . . . the ratio of income to investment can be calculated exactly from the [formula, using a few numbers available through routine accounting functions]. It need not be estimated from any model. The formula does not require empirical validation, and incorporates no empirical experience or wisdom. It is an accounting identity.
>
> The official . . . model ''explains'' [this accounting ratio] in terms of . . . a large number of ''explanatory variables.'' The model does not include [the exact formula] but it includes many expressions which are *more* complicated. If that particular ''interaction'' term had been included, the model would have accounted for 100% of the variation in profitability from one business to another, and not merely 78%.

If we have a shortage, it is a shortage of Ockham razors, not computing capability.

12
Affirmative Action and Discrimination

He who knows but little speaks the loudest.

Quechua Proverb

Facts do not cease to exist because they are ignored.

Aldous Huxley

I. Affirmative Action

We are using the term "affirmative action" to embrace all types of job discrimination based on sex, race, religion, age, or any other personal characteristic. The term "job" includes initial hiring, promotion, salary, discharges, etc. As a result of federal Civil Rights legislation various Executive Orders, guidelines, and other edicts have been issued. States and cities have followed suit and at present there are an untold number of proclamations. We cannot hope to cover even a small number of them but shall simply illustrate how they can and do result in the misuse of statistics.

We cannot comment on changes which occur after the completion of this manuscript. But for our purpose, which is to illustrate statistical misuses, the examples we give are useful. Keep in mind, however, that affirmative action practice may be different when you read this chapter. It is not our intent to make a political statement.

Perhaps the proclamation most widely known is the federal Executive Order No. 11246, and we begin with comments about it. All contractors to the Federal government had to devise affirmative action plans to assure that positive, affirmative steps were taken to guarantee equal employment opportunities for minorities and women. If a contractor did not have an acceptable plan the government could deny new contracts or revoke existing ones.

The concept is simple. For example, if women are 40% of the labor force in a particular occupation in the labor market area from which the contractor hires, then 40% of the workers in that occupation or among the new hires ought to be women. But the use of statistics to implement this concept is difficult.

We find the failure to understand statistical data and methodology, and a lack of knowledge of the dynamics of the labor market, rampant among the enforcers of Executive Order No. 11246. The implementation of this order was based on regulations written by individuals who did not know their subject matter.* We are not questioning the merits of equal employment opportunities or affirmative action as a principle. We are questioning the misuse of statistical methods to evaluate hiring practices, not only because such a misuse is misleading, but because we believe this misuse actually hindered the achievement of these goals which we believe are worthwhile.

A. What Are Affirmative Action Plans?

As required by the U.S. government, an affirmative action plan submitted by a contractor can be divided into three parts:

Employment procedures: The specific procedures he will follow to avoid discrimination against the "protected groups" of women and minorities† in hiring, promotion, salary, and personnel matters.

Utilization analysis: A statistical comparison of the sex and minority composition of the contractor's employees and new hires, as compared to the sex and minority composition of the labor market from which the contractor hires.

Corrective action: Action to be taken to bring the contractor's personnel composition in line with that of its labor market (if it is not in line) [3].

The first item, employment procedures, is nonstatistical, and we do not discuss it further.

---

*Our discussion of affirmative action is limited to the groups and processes covered by Executive Order No. 11246. In Chapter 11, Section II, we discussed the nonsensical statistical procedures used by the American Anthropological Association. We do not discuss the procedures introduced by states, other political entities, and private institutions.

†A person belongs to a minority if that person is a member of one of these groups: Black, not of Hispanic origin; Hispanic, a person of Spanish culture or origin regardless of race; Asian (excluding Southwest Asia but including the Indian sub-continent) and Pacific Islanders; Native American ("American Indian") or Alaskan Native.

The second item, the utilization analysis, is completely statistical and is the hornet's nest of misuses. On a clear day, it appears to the statistician with good vision that the federal government's rules for performing utilization analyses were written by attorneys unfamiliar with statistical procedures and that the rules were enforced by an equally (or less) informed investigative staff. Such a monumental lack of knowledge of the subject matter is a misuse we have discussed in Chapter 3, but can be forgiven.* However, if the attorneys who wrote the rules and the investigators who enforced them were statistically well-informed, then we would have to conclude that their misuse was deliberate, which cannot be forgiven.

The third item, corrective action, is a combination of statistics and negotiation between the government and the contractor. Our discussion of the statistical misuses found in the preparation of the utilization analysis applies equally to the "corrective action" phase.

B. Faulty Data

When making utilization analysis, the major source of data for the labor market is the U.S. Bureau of the Census.† In some cases, the data about the labor market have been obtained from other sources. The source of information about the contractor's personnel is invariably obtained from the contractor. However, all of the data, whether from the contractor, the U.S. Census, or elsewhere, can be deficient to a greater or lesser extent depending on many factors. Errors in the data will be reflected in uncertainty in the results obtained from those data. To make precise statistical conclusions from imprecise data is itself a misuse of statistics.

Even if precise data are used in a utilization analysis which suggests underutilization, you cannot say for certain that the firm did or did not engage in improper personnel practices (from the standpoint of utilization, not legality). All that you can conclude in the event of an apparent negative findng is that there is *evidence* of underutilization, which may be due to faulty statistics or the contractor's personnel practices, or both.

C. Problems in Defining Protected Groups

As we have discussed in Chapter 4, there are serious difficulties in identifying minorities. In the 1980 Census of the U.S. population, respondents were asked to

---

*If the misuse was the result of a lack of knowledge, it is no less a sin, because the staff could have consulted with experts who are knowledgeable.

†The appropriate source for a particular contractor depends on the labor market information concerning that contractor. The 1980 Census of Population information is available on computer tapes and in published reports. Both computer tapes and printed reports are available from, or can be located through, the U.S. Bureau of the Census, Washington, D.C.

identify their race/ethnicity. After the final checking was completed, the final census count showed almost six million people whose race was unknown [4]. If most of the unknowns were from minority groups, then about 12% of the minority population of approximately 50 million were not reported as such in the 1980 census.

The contractor has a unique difficulty in determining the firm's minority count of its new hires—a "Catch 22" for our time. This is because the law forbids an employer to ask an applicant about his or her race! Thus the contractor must guess at the race/ethnicity of the applicant by physical observation and examination of the name. Do you think this can be done accurately? In any event, once an applicant has been hired, race/ethnicity can be ascertained and entered in the contractor's records.

The major purpose of this law is to prevent employers from discriminating in hiring. But alas, affirmative action regulations require that the contractor also keep data on the race/ethnicity of *applicants* as well as employees to show what proportion of protected (including minority, of course) applicants were hired compared with the proportion of unprotected applicants who were hired.

But the potential employer often does not see the applicant, for application and rejection is frequently made through written or telephone correspondence. Assume the contractor knows the proportion of minority persons in his region. If he cannot ask an applicant his race and he does not personally interview the applicant, how can he ensure that he is in conformity? The applicant's name, address, or telephone voice may be the only evidence the contractor has, and this is surely questionable data!

D. Occupation

Utilization analysis rules require that contractors analyze the utilization of women and minorities separately for each occupation or group of occupations. How likely is it that a contractor has used the same classifications as the U.S. Bureau of the Census? Expert occupational analysts are needed to perform such job classification, and few contractors can afford to hire such a person. There are consultants in this field, but their fees are prohibitive for many companies. Thus the result of comparing the contractors' apples to the census's oranges often is what you might expect—nonsense.

There is also the effect of variability resulting from the process of sampling (as discussed in Chapter 2 and elsewhere), since occupation was asked only of a sample of all respondents in the 1980 Census. There were some problems in collecting and processing answers. These problems have been discussed in publications of the U.S. Bureau of the Census [4, see Appendixes].

Thus, the government's investigators misuse statistics when they demand that a contractor's occupational distribution and new hires by sex and race/ethnicity, match that of the labor market as reported by the Bureau of the Census or similar sources. And yet, if the occupational distribution and new hires contain fewer women and minorities than supposedly exist in the labor market, then the contractor is deemed to be "underutilized" and can be penalized. Or, if the firm has an apparent overutilization ("concentration") of protected groups in a given part of its labor force, then the contractor may be deemed to be guilty of "channelling" protected individuals into segregated occupations.

E. Seeing Double

Federal procedures treat women and minorities as separate groups, even though a woman can also be a member of a minority. A person can, simultaneously, belong to several categories. White non-Hispanic men are "unprotected." White non-Hispanic women, and both sexes of the minority population are "protected." These relationships are shown in Table 12.1. By counting women *and* minorities as two protected groups, we have double counting, bad arithmetic, and a statistical misuse.

For example, suppose that there are 1000 people in the labor force in a given area. There are 600 women and 400 men, and 800 white non-Hispanics and 200 members of minorities, giving the distribution shown in Table 12.2.

Table 12.1 Protected and Unprotected Groups

Ethnic category	Men	Women
White non-Hispanic	Unprotected	Protected
Minorities	Protected	Protected

Table 12.2 An Example of Counts of Protected and Unprotected Groups

Ethnic category	Men	Women	Totals
White non-Hispanic	a	a	800
Minorities	a	a	200
Totals (by sex)	400	600	1000

[a]Information not supplied.

If you count as the enforcers of the executive order do, you add the protected group of women (600) to the protected group of minority people (200), to get a total of 800 protected individuals in this labor force.

To get the correct total, you must know how many of the minority members are women. If we assume that 120 of the women are also members of minorities, then the table is as shown in Table 12.3. From this table you can see that the correct number of protected individuals is 80 plus 120 plus 480, adding up to 680. It is by double counting the 120 minority women that the enforcers get 680 plus 120, or 800.

The 1980 Census documents the existence of considerable overlap between women and minorities. In 1980, there were 45 million women in the labor force of, which about 9.5 million were minority women, close to one-fifth of all women in the labor force [5, Tables 86 and 134]. These 9.5 million women can be counted twice for a utilization analysis, as shown above. Thus, by hiring one minority woman, a contractor can gain credit for two protected hires.

We now show how this two-fold classification error can allow violations of the *intent* of the federal procedures while following all those procedures. Table 12.4 shows a hypothetical case.

Table 12.3 Completing the Counts of Protected and Unprotected Groups

Ethnic category	Men	Women	Totals
White non-Hispanic	320	480	800
Minorities	80	120	200
Totals	400	600	1000

Table 12.4 A Hypothetical Utilization Analysis

		White Non-Hispanic		Minority			
	Total	Men	Women	Men	Women	Women	Minority
Availability	100%	30%	30%	20%	20%	50%	40%
Contractor *X*'s personnel	100%	35%	25%	15%	25%	50%	40%

The utilization analysis of Table 12.4 shows that the labor market from which this contractor recruits has 50% women and 40% minority. Contractor *X* has these stipulated percentages and meets the federal requirements. But Contractor *X* has 5% *fewer* white non-Hispanic women, 5% *more* white non-Hispanic men, 5% *more* minority women, and 5% *fewer* minority men. All of this is done by hiring a few more minority women so as to benefit from the double counting.

E. Chance Factors in the Utilization Analysis

Chance factors may give the appearance of underutilization, or work in the opposite direction and cover up true underutilization. If the designers of the utilization analysis procedures had understood the dynamics of the labor market, they would have known that a contractor hires from among those who apply for jobs when the contractor is hiring. Of those who are hired, some who are found unacceptable lose employment with the contractor, others may leave for personal reasons such as getting a better job elsewhere. These are chance effects with important consequences.

Thus, at a particular time, a contractor may have long-term employees who do not fit the sex–race/ethnicity distribution of the labor market at the time of the utilization analysis. Are we to ask the contractors to fire some of these employees to please the investigators?

The appearance of a state of underutilization should be based on a consideration of the precision intervals which result from chance effects. If the contractor appears to be underutilizing with the chance effects taken into account, then the next step should be to examine the nonstatistical procedures which might be discriminatory and causing underutilization. To the best of our knowledge, the government's investigators still do not allow the use of the sampling variation in interpreting the utilization analysis.

New employees are hired from time to time. But the hiring is done from those who apply. Whether the new hires are in accordance with the labor market sex–race/ethnicity profile at the time of hiring is partly a matter of chance. Perhaps the jobs are for a "graveyard," or night, shift, for which women (or others) do not want to apply. Or perhaps the jobs are known to be temporary or dead-end jobs and the minority people in the labor market are seeking full-time jobs or jobs with potential for growth. There are many reasons why applicants are not necessarily a proper cross section of the labor market.

II. Discrimination

In an address to the American Statistical Association, Charles Mann summarized some of the abuses of statistics he saw in courtroom affirmative action cases [6].

We use three of his cases as the basis for examples to show both the misuse and proper use of "hypothesis testing" based on elementary probability theory.

A. Too Little Statistical Discrimination?

The first example we consider is: "In a particular company two of three under forty year old employees and one of two over forty year old employees receive raises.* Percentage [sic] are presented to a jury with the claim that a disparity in treatment is clear [6, p. 3]."

First, let's compute the percentages. Of the five company employees, two are over forty years of age ("over forty"); the percentage of over-forty employees is 40%. Of the three employees who received raises, only one is over forty; the percentage of over-forty employees getting raises is about 33%.

The inference to be drawn from Mann's statement, as presented to the jury, is that the difference between the percentage of over-forty employees receiving raises (33%) and the percentage in the labor force of interest (40%) is great enough to indicate age discrimination against over-forty employees. Yes, 33% is less than 40%. But what about chance effects? To take them into account, we state the issue this way: If there were no age discrimination, how likely is it that a percentage as low, or lower, than 33% would have been observed? We can "test" for discrimination by taking as our null hypothesis the assumption that there is no age discrimination and seeing if the probability of getting 33% or less *by chance* is low enough to give us reason to reject the null hypothesis.

We cannot have absolute certainty. Even if the null hypothesis were true and no age discrimination was in effect, chance might lead to an observed percentage of 33% or less; this is the probability we must calculate. But if the probability of getting this result by chance is very small, then we are inclined to reject the null hypothesis, and we have a basis for arguing that age discrimination exists.† If the choices of individuals for raises has no relationship to their ages, then the probability of picking an over-forty employee for a raise is .4. You can use this analogy. Suppose you have a bag with five apples in it, three red and two green. If you randomly (apples suitably mixed, no peeking in the bag) choose one apple, what is the probability of picking a green one? Since 40% of the

---

*We assume that Mann means "age forty years or less" and "over age forty years."

†If this short and elementary summary of the rationale of hypothesis testing troubles you, consider this analogy: Your dog has eaten your neighbor's nasturtiums. You want to investigate your neighbor's feelings toward you after this incident, and approach the neighbor while she is watering her garden. Your null hypothesis is "neighbor bears no resentment toward me." The neighbor turns, faces you, and sprays you with water from the garden hose. How likely is this occurrence if the null hypothesis were true? Rather unlikely, so you reject the null hypothesis that the neighbor bears no resentment. On the other hand, you can't be certain. The neighbor might have been startled and turned the hose toward you by chance. The probabilities determine your conclusions.

apples are green, the answer is .4. It is the same for picking over-forty employees for raises, *if* there is no age bias in the choice. The statistical formulation of the test of this null hypothesis (of no age discrimination) is this: If you choose three people out of a group of five for raises, without concern for their ages, what is the probability that one or none of them will be over forty, given that two of the five are over forty? The answer to this question is the computed probability of .7.*

To summarize, the probability of giving raises to none or one over-forty employee, out of three chosen at random, is .7; this is an extremely likely occurrence! Conclusion: It is very likely that this could have occurred by chance. Given the assumptions and this probability, there is no basis for saying that age discrimination was at work. This doesn't mean that there was no age discrimination, only that the numbers reported do not give you a basis for saying that there *was* age discrimination. To use the data of this case to claim age discrimination is a misuse of statistics, plain and simple.

B. Divide and Conquer—Something, But What?

Another example which we analyze with the principles of the preceding section is:

> After properly applying a statistical test [that is, a hypothesis test] for disparate treatment to fifty independent corporate locations it is noted that at two locations statistical significance (at the .05 level) is obtained. These two locations are isolated and their data presented as indicative of "discriminatory" behavior on the part of the company. [6, p. 3]

A test of some hypothesis is said to have *statistical significance* if the probability of the observed result occurring by chance is less than some specified probability, called the *level of significance*.† In the social sciences, most researchers use a probability of .05 as the probability which determines statistical significance. In the example of the prior section, the probability was .7 of choosing one or no over-forty employees, if the null hypothesis of no discrimination was true. This value is much higher than .05, and we say that the observed

---

*This is the sum of the probabilities for getting one over-forty employee and no over-forty employee. Expressed using the binomial coefficient $C(N,r)$, the number of combinations of N objects taken r at a time, this probability is $C(2,1)C(3,2)/C(5,3) + C(2,0)C(3,3)/C(5,3) = .7$.

†"A procedure which details how a sample is to be inspected so that we may conclude that it either agrees reasonably with the hypothesis or does not agree with the hypothesis will be called a *test* of the hypothesis; it is a decision rule which tells us to *accept* the hypothesis for certain types of samples and *reject* it for other types [7, p. 76]. . . . The values of the statistic [the observed result] . . . which we agree should cause rejection of a hypothesis will be called a *critical region* for the test, and the probability that the statistic [the observed result] . . . will be in the critical region when the hypothesis is true will be called the *level of significance*" [7, p. 82].

results were not statistically significant. If the probability of choosing one or no over-forty employees by chance was less than .05, we would have said that the test had statistical significance. This means that we would reject the null hypothesis, and it would be reasonable to assume that age discrimination is taking place.

In this case, we have 50 locations of a corporation, in each of which a hypothesis test for discrimination (''disparate treatment'') was used with ''no discrimination'' as the null hypothesis. For two of those locations, the probability of the observed result occurring by chance if the null hypothesis of no discrimination were true was .05 or less. Thus, it would not be unreasonable to conclude that discrimination could be occurring *at these two locations*. But can these two out of 50 locations be used to condemn the whole corporation of discriminatory behavior?

To answer this question, we must perform a test on the corporation as a whole. The null hypothesis is that the corporation does not discriminate. Divide the corporation into 50 separate locations. Apply a test at each location *which has a probability of .05 of showing discrimination by chance, even if it doesn't exist.* Thus, assuming that the null hypothesis is true (no discrimination), what is the probability that two or more of these 50 tests will be statistically significant at the .05 level?

We use this rationale to compute the probability. Each location is an independent ''trial'' which has a probability of .05 of being found to be statistically significant; but we want to know the probability that two or more locations will be statistically significant in 50 trials. We calculate the probability of getting two or more significant results out of 50 to be .26.* Thus, the observed result could easily have occurred by chance; the case against the company is not proven. You might ask: In this situation, how many locations must show a statistically significant result for us to reject the null hypothesis that the company is not discriminating? The answer is: five locations.

If you infrequently compute probabilities, or have not had a course in statistics, you may feel that this is abtruse reasoning and that these are difficult computations. However, setting up hypothesis tests and finding the probabilities in the two preceding cases are typical homework and examination questions students must solve in statistics or quantitative analysis courses. These methods should not have been beyond the capabilities of attorneys, judges, their advisers, and especially those investigators who implemented Executive Order Number 11246 and other Civil Rights rules.

---

* We used the binomial probability distribution.

C. And Now *for* the Other Side of the Coin

Sometimes a conclusion of "no evidence of discrimination" is not justified by the data. We illustrate with a simplified version of one of Mann's cases. Five candidates were considered in each of 24 independent promotion decisions. In every decision, there were two females among the candidates, and in every one of the 24 decisions, a male was promoted. The statistical misusers performed an individual hypothesis test on each of the 24 decisions. In every case, the test was not statistically significant, and the null hypothesis of no discrimination was not rejected. Thus the analysts concluded there was no discrimination in each *individual* decision.

We can easily describe the methods that were used and those which should have been used. The null hypothesis is that there is no sex discrimination, and obtaining a male as the successful candidate is simply the probability of randomly getting a male on one draw out of a box containing three males and two females. This probability is 60% (three out of five). This is a likely occurrence and gives no basis for rejecting the null hypothesis of no discrimination.

However, these cases must be looked at as a whole as well as individually. What is the probability that no females will be chosen in 24 decisions, in the same organization where the probability of drawing a female is .4? We get this probability from the binomial distribution, it is .000005. This is such a small probability that it is almost impossible (not just unlikely) that the observed result could have occurred by chance if there was no sex discrimination at work.

Alas, in the actual case from which this example was derived, "Since no individual case allows a conclusion of disparity (at the .05 level), it is concluded that there is no reason to believe that promotions are disproportionately awarded to males." So much for achieving the intent of the executive order when the dead hand of bureaucracy misuses statistics.

Sometimes it is reasonable to look at individual cases; in others, it is not. Robert Hooke, in *How to Tell the Liars from the Statisticians,* gives an example in which the whole organization appears to discriminate in their hiring decisions, although there is no discrimination in the individual decisions [8, pp. 26–7].

III. Summary

We find at least five statistical misuses in the affirmative action and civil rights procedures we have discussed:

Lack of knowledge of subject matter

Improper design of analytical procedures

Use of faulty data

Failure to allow for chance or sampling variation

Improper use of hypothesis tests and probability computations

Any one of these misuses is enough to indicate the absurdity of the procedures in which they are used. Are these misuses permissible because the overall utilization program is socially desirable? We believe that affirmative action *is* a socially desirable program, but social desirability is no excuse for misusing statistics and then claiming that "statistics prove the case." In the long run, good will and the proper use of statistics are the tools that will achieve the program's goals.

13 Ectoplastistics

ec′tō·plasm, *n.* [*ecto-*, and *-plasm.*] . . . 2. in spiritualism, the vaporous, luminous substance supposed to emanate from the medium's body during a trance.

Webster's New Twentieth Century Dictionary

I. Introduction

Spiders and humans have an ability in common: spiders emit silk from their bodies to catch unwary insects; humans emit thoughts—both myths and statistics—with which to trap the unwary hearer. Some of these human emanations are so "vaporous" and "luminous" that they are accepted by the unwary as "the truth."

Is the real difference between *homo sapiens sapiens* (to distinguish ourselves from long-gone ancestors) and its predecessors the ability to produce numbers? If so, it is both a great advantage and disadvantage, for numerical emanations are often emitted without serious regard for their truthfulness. Thus are created **ectoplastistics** [n., *ecto-, -pla* and *-s(ta)tistics*], which can be used to get legislation passed, raise armies, create antagonism among peoples, extol or slander, win political position, change economic activity, and so forth.

No human activity is immune from the influence of ectoplastistics. Some readers may remember, as do the authors, that children of the 20s and early 30s were taught that they must chew a mouthful of food exactly 32 times before swallowing. Thirty-two times—no more, no less. Would any other number have been as effective? In this case "32" was pure ectoplastistics, an emanated number based on a spiritual substance with no more reality than being what some authority imagined (or wanted) to be the truth.

In this chapter we review more recent examples of ectoplastistic numbers.

II. Negotiation Is Honest, Ectoplastistics Is Fake Truth

What is the worth of a person? Many decades ago, newspapers made headlines by reporting the estimated value of a person, which was arrived at by summing the average values of the chemical components of the human body. Average human values were used and neither sex nor race entered into the calculation. At the time, the value was less than one dollar.

More recently, *comparable worth*, a new method for determining the value of a person, has made headlines. It asks: Should all people who work in jobs having an equal worth be paid the same wages? ". . . jobs that are equal in their value to the organization ought to be equally compensated whether or not the work content of those jobs is similar" [1, p. ix].

We agree with this concept and imagine that most people who work (which is most of us) agree as well. But how are we to determine which jobs are comparable in their value to the organization? Is the value of a secretary comparable to that of a plumber? Is the value of a plumber comparable to an electrician? Is a photographer worth as much as a fashion model? Is the value of either the photographer or the model the same, less than, or more than that of the designer of the clothes being modeled? Is a painter working with paints worth the same as a painter with the same skills, adding the same value to the product, but working in a hazardous environment? If you agree with the basic concept, then you must face the question: *How do you decide on equality of value* (*comparable worth*)?

Differences in pay between individuals may well have existed since work has been paid for. The basis for differences may be the type of employment, the sex or age of the worker, or even geographical, as in the wages of hand laborers in two different regions of France in the 16th Century [2, p. 303]. In more recent times, unionized construction workers in New Jersey in 1898 earned the following daily rates [3, p. 580]:

Job	Median Daily Rate (dollars)
Bricklayers	3.80
Plumbers	3.00
Carpenters	2.53
Painters	2.50

What's the story? Were bricklayers worth more to the construction company than plumbers, carpenters, and painters? Or were there just fewer bricklayers available in the labor pool?

Differences between men's and women's wages, and adult's and children's wages, for the same work are well documented in the United States. The U.S. Bureau of Labor conducted a survey in 1895–1896 to cover "specifically the employment and wages of women and children in comparison with the employment of men in like occupations" [4, p. 30].

The Bureau surveyed thousands of establishments and collected data for over 150,000 workers. It published information on occupations, hours worked, and comparative earnings of men, women, and children "of the same grades of efficiency." The finding was that "for the same occupation and grade of efficiency, men earned over 50 percent more than women, and that children earned substantially less than adult workers" [4, p. 31].

Why do women get about two-thirds of men's wages for the same jobs at the same "grade of efficiency"? We find that this question recurs again and again.

Returning once more to New Jersey in 1898 for a specific case, we find that the silk industry paid *men* the following rates [3, p. 579]:

Job	Median Weekly Rate (dollars)
Weaving	10.53
Dyeing	9.73
Throwing	5.33

Once again we could ask about the value of these jobs to the company. But instead, let's look at what women were paid in the same industry, location, *and in the same jobs*. The median weekly rate for women in weaving was $7.00. Could women in weaving really be worth only two-thirds as much to the employer as men in weaving?

In Massachusetts in 1893, the median weekly wages for men employed in manufacturing was $10.00, and for women $6.40 [5, p. 288]. Is it possible that women employees were worth only two-thirds as much as men employees on the average? Or, is this the free market at work? In 1888, the Bureau of Labor (the agency now known as the Bureau of Labor Statistics) issued "Working Women in Large Cities," based on a national survey. Many women agents worked in the field to collect data, and the report notes that "They have stood on an equality in all respects with the male force of the Department, and have been compensated equally with them." But this was the report quoted above, which found that men earned 50% more than women of the same "grade of efficiency," and the commissioner of the Bureau noted that "women were willing to work for lower wages than men" [4, p. 30].

In the 1980s, the discovery that women earned less than men made headlines and political ferment. Buttons with the slogan "59 cents" were worn to indicate that the average pay of women was 59% of men's pay. Fifty-nine percent! Allowing for the variability in statistical measurements due to data, differences in computational method, and errors, this ratio has hardly changed in almost 100 years!

Many factors enter into the comparable worth equation. A major factor is that male and female workers tend to concentrate in different jobs, and there is no question that the jobs in which women predominate pay significantly less than those in which men predominate. Thus:

> For many women, the slogan "Equal pay for work of equal value" has replaced the slogan "Equal pay for equal work" which is embodied in the Equal Pay Act of 1963. More generally, the issue raised is that of pay equity in a labor market which is highly segregated by sex. [1, p. 2]

We illustrate this point by comparing two health care professions, pharmacy and nursing. In 1979, 84% of the pharmacists in the United States were men, and the median earnings for year-round workers were $23,000. In the same year, 95% of the nurses were women whose median earnings were $15,000 for full-time employment [6, Table 281]. Are pharmacists more valuable to the health maintenance of the people of the United States? Are nurses really only worth 68% as much as pharmacists? Or does the work of the pharmacist have the same value as that of the nurse? If this is the case, then women are simply not getting equal pay for work of equal value.

On the average, do women get the same earnings as men when they work in the *male*-dominated profession of pharmacy? Not if we believe the 1980 U.S. Census data. The median wage for full-time male pharmacists is about $23,000; for females about $18,000. The median female earnings in pharmacy are 78% of the median male earnings [6, Table 281].

Do women get the same average earnings as men when they work in the *female*-dominated profession of nursing? From the same source, we find that the median wage for full-time male nurses is about $16,000; for females about $15,000. The median female earnings in nursing are 94% of the median male earnings. [6, Table 281].

Sincere proponents of equal pay for equal worth could argue that the jobs of pharmacists and nurses—or any other set of jobs—could be evaluated for comparability of value, and those which are of equal value to the employer or society should receive equal pay. If the analysis indicates that nurses and pharmacists are of equal value, then the median wages of nurses should be increased to those of pharmacists, or the median wages of pharmacists should be reduced to those of nurses, or both should be moved to meet somewhere in between. People being people, it is almost impossible to reduce wages of one group to create wage equity for another group. Thus, making an inequitable wage distribution equitable is almost certain to lead to higher wages for some workers and, hence, increased overall costs. This fact may have more to do with the resistance to the implementation of comparable worth concepts than sexual (or other) discrimination.

In any event, the comparable worth concept reduces to the determination of "equal worth." How to do this? The reflex answer is: with statistics, of course. Everyone knows that statistics have magical properties. Statistics can and will prove the case once and for all. But which statistics do you use? How do you process them to get "correct" answers? And we must still question whether statistical analysis is the way to do it.

The use of statistics to ascertain the worth of jobs predates the present controversy, and there are many different methods in use. Almost every firm or organization which uses statistics this way has a particular method customized to its own needs and situation. We cannot review all the major approaches, so we've settled for describing the generalized procedures. The methods we have selected are not used in all cases, or may be totally rejected or rigidly enforced in others.

Most such attempts start by preparing job descriptions. You can search for reliable sources of standards for job descriptions to use as starting points such as textbooks, compensation consultants, corporate policy statements, and procedures. But most final versions are *ad hoc* responses to a given situation or the desires of a particular executive, manager, supervisor, or administrator or the result of a consensus among interested parties.* In general, a job description includes a description of the nature of the job, the skills, experience, amount of effort required, the responsibility and authority of the job, the level of formal education (or its equivalent) required, and personal attributes of the individual (where important). The description should quantify the requirements and attributes which form the basis for determining the job's "value." For example, a systems analyst, in addition to possessing certain technical knowledge, has to work well with people from different functions and at different levels. The ability to carry out this task without arousing hostility is vital to most such jobs; but this attribute, like most others, is difficult to quantify.

Anyone who has prepared any job descriptions (as we have) knows the difficulties of making the job description and the actual work conform. In many organizations, there are no standard lists of aspects or necessary elements and no generally accepted standard elements except for the most elementary tasks. Each firm (and even each subunit within the firm) may decide on the job elements. This makes it difficult to compare jobs among different firms or even within a firm. And the personal characteristics of the individual who was hired to fit a job

---

*For an overview, see Ref. 1, p. 95.

description often creates the worth the organization assigns to that job. As it is carried out, the job is as much a function of the individual's ability and the individual's changing role in the organization as it is of any formal preliminary description. Even at the level of clerical and manual labor, individual ability, preference, and on-site needs often determine what is actually done.

Beyond the initial job descriptions, there are other formal aspects of the job evaluation process. Each element of the job description may be given a relative value, depending on its importance as judged by the members of the firm or the specialist preparing the job description. For example, if the organization has a job in which a college degree is considered important, it may give the college degree a weight of 3. If a college degree is desirable but not important in another job, the college degree may get a weight of 1. This type of analysis is often quite useful in comparing candidates (as opposed to comparing jobs).

But how useful is it in comparing jobs? And how might it be used to compare jobs? Let's look at the spiritualistic process of producing high-quality ectoplastistics. Table 13.1 shows some hypothetical ratings which were produced by the subjective weighting process just described [1, p. 121].

In the process of producing ectoplastistics, the vertical columns are inspected for similar patterns; jobs which have the same patterns are declared to be of equal worth. As McCormick stated in his minority report [1, p. 115 ff.],

> Jobs A and B are identical, job C is almost the same as jobs A and B, but job D differs markedly from the other three. The fact that data for the job elements are quantified makes it possible to compare jobs in quantitative terms. Most typically, some statistical index of similarity is derived for each pair of jobs. In turn, such indexes frequently are used for grouping jobs into groups that have reasonably similar profiles of job element values. [1, p. 121]

Despite the use of "statistical index(es) of similarity," the decision as to whether jobs are similar is judgmental. The analytical nature of the index does not override the overwhelmingly judgmental nature of the raw data on which the entire comparison and analysis rests. Subjective judgments are subjective judgments, and you do not have to be sophisticated in statistics (or the writing of job descriptions) to see the tenuous relationship between the performance of the job (and its value) and the vertical columns of ectoplastistic numbers. "The process is inherently judgmental and its success in generating a wage structure that is deemed equitable depends on achieving a consensus about factors and their weights among employers and employees" [1, p. 96]. In short, even the supporters of the process acknowledge that the basis for calculating worth is pure ectoplastistic emanation.

In the effort to legitimize the process by appearing to get around the judgmental aspects, the analysts may use regression equations, concordance indexes, and other complex statistical procedures. But these hairy procedures

Table 13.1 Hypothetical Job Description Patterns

Job Element	Job A	Job B	Job C	Job D
a	1	1	1	3
b	4	4	3	1
c	0	0	0	2
d	3	3	3	0
-	-	-	-	-
n	2	2	2	4

Note: Ratings: 0 = low; 5 = high.
Source: Ref. 1, p. 121.

almost invariably need shaves with Ockham's razor (Chapter 11). And none of these highfalutin' procedures can get around the basic questions: How accurate are the job descriptions in describing the work to be performed? Have all the proper factors (elements) of the job been specified? Are all the specified elements relevant to the job performance? Do the weights accurately reflect the role of the element in job performance? Does the content of the job vary with the capabilities of the individual worker? Hairy, and even well-shaven statistical procedures may only be a disguise, camouflaging the subjective and imprecise judgments which give rise to the basic data. Therein lies the misuse of statistics.

By the mid-1980s, the terms "equal worth" and "comparable worth" began giving way to the term "pay equity." In June 1985, the Equal Employment Opportunity Commission ruled unanimously that equal pay for jobs of comparable worth was not required by federal law [7]. However, suits were still being heard on variations of this principle.

More importantly, in our view, employers and unions (and in some cases, groups of nonunion employees) are *negotiating* wage increases for the lower-paying jobs mostly held by women (nurses, librarians, secretaries, and so forth). Many of these negotiated settlements are between state and local governments and governmental employee unions. As far as we can judge from newspaper reports (we have no first-hand evidence, and it is too early to have objective data), the use of statistics, beyond summaries of current wage rates, is a minor element [8, 9]. As reported, this process is based on an explicit recognition of the fact that there are low-paying jobs held mostly by women, and in light of the extolled virtues of pay equity, wages are being raised by mutual agreement. To achieve this, no complex statistical methods need be run on computers to massage data of dubious value, thereby creating another misuse of statistics. We conclude that these open negotiations are a far more honest approach to achieving equitable pay for workers than ectoplastistic analysis.

III. Just How Much Does Addiction Cost?

Around 1970, it was commonly believed that New York's heroin addicts committed about half of all property crimes. Many prominent individuals and organizations supported this view, which stated that the amount of property theft by heroin addicts was between two and five billion dollars per year. Is this an ectoplastistic emanation or an estimate with a reasonable basis in fact? According to Max Singer, the estimators made assumptions and worked as follows:

> There are 100,000 addicts with an average habit of $30.00 per day. This means addicts must have some $1.1 billion a year to pay for their heroin (100,000 × 365 × $30.00). Because the addict must sell the property he steals to a fence for only about a quarter of its value, or less, addicts must steal some $4 or $5 billion a year to pay for their heroin. [10, p. 3]

How does this reasoning stand up? If we accept the assumption that the total number of addicts is 100,000, how many addicts steal? Some of the addicts can pay for all or part of their habit without stealing. In addition, there are data which support the premise that the street addicts who steal spend about one-quarter of their "careers" as addicts in jail. This reduces the estimate of the number of addicts at any one time to about 75,000, according to Singer.

The figure of $30.00 a day is based on the amount of heroin consumed and its street price. Is it reasonable? An unknown number of heroin addicts are in the trade and get their heroin at a wholesale cost significantly less than $30.00 per day. Addicts steal cash as well as property, which means that they sometimes can get full value for their thefts without going through a fence. These are unknowns, so other methods of valuation must be used. Singer asked, "What happens if you approach the question from the other side? 'How much property is stolen—by addicts or anyone else?' Addict theft must be less than total theft. What is the value of property stolen in New York City in any year?" [10, p. 4].

Singer could not get any supported estimate of total theft. He therefore made his own estimate based on fragments of information available, which we do not repeat, but which you can judge by reference to his article. His conclusion:

> If we credit addicts with all of the shoplifting, all of the theft from homes, and all of the theft from persons, total property stolen by addicts in a year in New York City amounts to some 330 million dollars. You can throw in all the "fudge factors" you want, and all the other miscellaneous crimes that addicts commit, but no matter what you do, it is difficult to find a basis for estimating that addicts steal over half a billion dollars a year, and a quarter billion looks like a better estimate, although perhaps on the high side. After all, there must be some thieves who are not addicts. [10, pp. 5–6]

Thus, the two to five billion dollar value is an enhanced ectoplastistic emanation, which vaporizes rapidly on examination. Singer gives good advice on how to examine ectoplastistic emanations: "The main point of this article may well be to illustrate how far one can go in bounding a problem by taking numbers seriously, seeing what they imply, checking various implications against each other and against general knowledge" [10, p. 6].

We agree with this approach, which is one of the ways by which both ordinary people and experts can avoid being led astray by ectoplastistic visions.

IV. Are the American People Becoming More Litigious?

"The courts in the United States are overburdened. Why? Because Americans litigate too much. They run to the courts at the slightest provocation, real or imaginary." This is the cry of some members of the judiciary, the most notable of whom is Chief Justice Burger of the U.S. Supreme Court. To prove the contention, statistics are quoted. Are the statistics real or imaginary? According to Marc Galantner, Professor of Law and South Asian Studies at the University of Wisconsin-Madison and president of the Law and Society Association, the statistical evidence usually given is:

> 1. The growth in filings in federal courts;
> 2. The growth in size of the legal profession;
> 3. Accounts of monster cases (such as the AT&T and IBM antitrust cases) and the vast amounts of resources consumed in such litigation;
> 4. Atrocity stories—that is, citation of cases that seem grotesque, petty or extravagant: A half-million dollar suit is filed by a woman against community officials because they forbid her to breast-feed her child at the community pool; a child sues his parents for "mal-parenting"; a disappointed suitor brings suit for being stood up on a date; rejected mistresses sue their former paramours; sports fans sue officials and management; and Indians claim vast tracts of land; and
> 5. War stories—that is, accounts of personal experience by business and other managers about how litigation impinges on their institutions, ties their hands, impairs efficiency, runs up costs, and so forth. [11, p. 12]

Even if this evidence indicates that we have a great deal of litigation, how do we know that it is too much? Do the growth and expenditure statistics (1, 2, and 3 above) and anecdotes (4 and 5 above) tell us that the cause of overburdened courts is that Americans litigate too much?

What can we say about usage of the law courts in America? There is statistical reason to think that the rate of litigation was much higher in the "good old days" of Colonial America:

> In Accomack County, Virginia, in 1639 the litigation rate of 240 per thousand was more than four times that in any contemporary American county for which we have data. In a seven year period, 20% of the adult population appeared in court five or

> more times as parties or witnesses. In Salem County, Massachusetts, about 11% of the adult males were involved in court conflicts during the year 1683. [M]ost men living there had some involvement with the court system and many of them appeared repeatedly. [11, p. 41]

Closer to our time, Galantner describes an analysis "of federal district court activity from 1900 to 1980 [which] shows a dramatic reduction in the duration of civil cases from about three and a half years at the beginning of the century to 1.16 years in 1980," and another which reports that "The number of cases terminated per judge has been steady since World War II and remains considerably lower than in the inter-war period" [11, p. 37].

For another viewpoint, consider the growth in the number of filings. Not all of the cases that are filed come before judges. Many filed cases are dropped or settled without the intervention of a judge and never reach the courtroom. The number of filings certainly is greater than the number of cases heard in the courts.

Galantner could find no consistent pattern in prior analyses and statistics to indicate a widespread large increase in the use of law courts. Unfortunately, the date he could find were limited, but some ideas can be derived from them. For example, a study of the St. Louis Circuit Court revealed that 31 cases per thousand of population were filed in the decade 1820–1829. By the decade 1890–1899, the rate had fallen to seven per thousand of population, but it rose to 17 in the period 1970–1977. In other localities for which records exist, the rate can be shown to have decreased over the last century, to have risen, or to simply have wandered up and down without any trend [11, pp. 38–41]. Is the United States an exceptionally litigious country? Galantner found

> . . . the United States rate of per capita use of the regular civil courts in 1975 was just below 44 per thousand. This is in the same range as England, Ontario, Australia, Denmark, New Zealand, somewhat higher than Germany or Sweden, and far higher than Japan, Spain, or Italy. . . . Given the serious problems of comparison, it would be foolhardy to draw any strong conclusions about the relative contentiousness or litigiousness of populations from these data.
>
> . . . The United States has many more lawyers than any other country— more than twice as many per capita as its closest rival. In contrast the number of judges is relatively small. The ratio of lawyers to judges in the United States is the one of [the] highest anywhere; the private sector of the law industry is very large relative to the public institutional sector. (Perhaps this has some connection with the feeling of extreme overload expressed by many American judges.) [11, p. 55]

The real issue, according to Galantner, may be the changing nature of the U.S. society over the decades and the changing nature of the work of the courts. With the passage of laws concerning environment, health, safety, and welfare, it is inevitable that cases involving these laws would come to court. Such cases may arouse more emotional response on the part of judicial representatives of the government than the large volume in colonial times of cases between individuals, since the current cases often involve the government itself.

For example, U.S. Social Security laws provide benefits for disabled people. For a period of several years in the early 1980s, the executive branch of the government chose to not follow the law and cut off these benefits. Thousands of disabled people went to the courts to force the government to follow the law as written. The courts ordered the government to obey the law and resume the payments. If the government's purpose was to get a court ruling on the law, only one court case would have been needed. However, for some time the government refused to follow the court rulings (including those of the U.S. Court of Appeals) and more and more cases were generated. Finally, on June 3, 1985, the executive branch announced that it would follow the law [12].

Despite this statement, the federal government continued to generate lawsuits through its own actions, resulting in action by a federal judge:

> A Federal judge in Manhattan yesterday [August 19, 1985] ordered the Reagan Administration to follow court precedents in determining who in New York State is eligible for Social Security payments.
>
> The Administration, which over the last four years has tried to cut off monthly benefits to almost half a million people throughout the country, has been restoring payments to individuals who win lawsuits against the Government. But it has refused to restore benefits to thousands of people in like situations who failed to bring suit.
>
> Last June . . . the Administration agreed to change its policy. It said it would apply the guidelines set by court rulings but only for benefit recipients who protested the initial denial of payments and appealed through an administrative process. [13]

How many thousands of unnecessary cases went to courts because the executive branch of the government refused to follow the law? How much of the heavy load perceived by Chief Justice Burger and the other members of the judiciary derives from sources such as this? We don't know, but discerning people, as well as plaintiffs, are entitled to regard the statements about the "crushing" court load as based on ectoplastistics, and act accordingly.

V. Ectoplastistics Do Not Nourish Live People

You can't eat ectoplastistics, but it can nourish the body politic. Governments have a big stake in publishing ectoplastistic information on agriculture. Under Haile Selassie, the prerevolutionary government of Ethiopia published a series of statistical reports which seemed to show that the government's bureaucrats knew how much food was being produced and how many Ethiopians there were to eat this food. But William Abraham found out that the government did not know the agricultural production level or the population count.

Abraham worked in Ethiopia from 1972 to 1975 as part of a team of planning experts advising the Ethiopian government on economic problems and future planning. His work led him to investigate a number of government reports containing statistical time series, including industrial production, agricultural production, foreign trade, and so forth. To know the existing state of affairs and to plan for the future, he needed statistical information. Alas, instead of information, he found ectoplastistics.

> Of the many gaps in economic intelligence the most serious by far proved to be the absence of reliable information on agriculture, a sector which accounts for around half the GDP [Gross Domestic Product] and provides a livelihood for something like 85 per cent of the country's population. This shortcoming is by no means obvious to the unwary since the statistical abstracts published by the CSO [Central Statistical Office] contain area, yield and production estimates of all major crops. It was only after much digging and detective work that I came to understand that the crop production figures were nothing but mechanical extrapolations of an old set of doubtful benchmark estimates. I was able to trace responsibility for the benchmark figures to a group of experts assembled by a planning minister many years ago and locked up in a room until they could agree on a set of estimates. This manful attempt to pierce the veil of mystery surrounding agricultural production was based on the group's best estimate of cereal consumption per head in kilograms (cereals form a very important item in the average Ethiopian's diet), the preliminary results of a sample survey of cropping patterns, some idea of yields and seeding and loss rates, and a bold guess of the population size. For succeeding years the base year production figures were simply extrapolated by the assumed rate of population growth. Since Ethiopia has still to take its first complete census, population growth had to be inferred from the results of scattered sample surveys and old counts of certain urban centers made by the Ministry of the Interior. I myself have never dug deeply into the subject of Ethiopia's population but I am aware that growth rates for the country that have found their way into print vary from about 2.0 to over 2.5 per cent annually, a very wide range where compounding has to be applied. This then, I learned to my dismay, was the basis of Ethiopia's series for grain production. [14]

Here is a graphic description of group production of ectoplastistics—emanations piled on emanations. The current government of Ethiopia is providing statistics on food production to the United Nations Food and Agricultural Organization, which dutifully publishes them annually. How accurate are they? Ectoplastistics, or reasonably factual? Who knows?

On May 9, 1984, Ethiopia took its first census of the population and published its unbelievably precise preliminary count of 42,019,418. How did the government get this figure? The official report tells us "Population estimate for areas that were not covered by the census is made using the official population estimate published by the Central Statistical Office" [15, p. 14]. But at the time of this census, a major famine was under way and hordes of the hungry and displaced were moving about (and out of) the country, which makes it very difficult to make accurate counts of people. Perhaps little has changed since the days of Haile Selassie, statistically speaking.

VI. Ectoplastistics in Court: How Many Workers Will Get Asbestos-Related Diseases?

Since about the time of World War I, it has been known that prolonged exposure to asbestos fibers can cause cancer and related diseases. But, as Paul Brodeur pointed out [16], many key organizations ignored warnings of these effects, including the federal government, Johns-Manville (the major manufacturer of asbestos products), and many firms that installed asbestos products. Apparently, thousands of workers who installed the asbestos products in buildings, factories, schools, naval vessels, and many other places did not know that inhalation of the asbestos fibers could cause asbestosis and the equally deadly mesothelioma. In the heyday of the asbestos industry, they endured these sicknesses largely in ignorance of their origin in the workplace.

Initially, a few court cases were filed by workers seeking compensation. The number of cases, as well as the amount of damages awarded by judges, juries, or in negotiated settlements, grew. By the late 1970s, as the largest producer of asbestos products, Johns-Manville found itself on the losing side of a large number of cases. Over and over again, it was proved in court that Johns-Manville had long known that asbestos fibers were dangerous to the health of workers and deliberately chose to ignore this information. Indeed, they even attempted to suppress it, but (even in that time before computer storage of data) written documents survived in corporate files. The plaintiffs' attorneys were able to introduce them as evidence.

What was the company to do in the face of this rising tide of increasingly expensive courtroom defeats? Company officials decided that voluntary Chapter 11 bankruptcy* was a good way to protect Johns-Manville from being forced to pay out "too much" money to victims and their attorneys. But to do this, Johns-Manville had to give the court estimates of the probable future claims from pending and future lawsuits that the company would have to pay. If the total of the estimated losses was judged by the court to be too great, or if no estimate could be made, then it was very likely that Johns-Manville would not be permitted to file a Chapter 11 bankruptcy.

Therein lies the motive for Johns-Manville's emanations of ectoplastistics. The company commissioned a study of the possible future incidence of asbestos-

---

*A Chapter 11 bankruptcy allows the company to survive and continue operations after meeting agreed-upon obligations, such as covering the costs of estimated future claims for compensation.

related cancer and the number of potential claims. An epidemiologist, Nancy Dreyer, was hired to do the job. She analyzed data supplied by the company and estimated a total of 230,000 additional cases of asbestos disease and 49,000 new lawsuits by the year 2000. This contrasted sharply with estimates made by Dr. Irving Selikoff, one of the world's experts on asbestos disease, who predicted 270,000 excess deaths from asbestos-related cancer alone.

The Dreyer report was honest about its uncertainties, saying that: "the actual number of lawsuits might easily be as low as half or as much as twice the number our calculations suggest" [16, July 1, 1985, p. 42].

The management of Johns-Manville was less than happy with these uncertainties which meant that this report could not be used in their request for a Chapter 11 bankruptcy. They then directed the team which made the report to "refine the estimates" and to try to lessen the range of uncertainty.

The company brought in Marc Victor, a legal decision analysis expert, to "assist in this refinement." Victor had neither experience nor training in epidemiology, but nevertheless he succeeded in

> . . . persuading Dr. Walker, a [public health] scientist, to reshape some of his assumptions concerning the future incidence of asbestos disease. . . . Walker revised his original estimate in such a way as to lower the projected number of people who might develop lung cancer as a result of exposure to asbestos . . . by discounting the risk multiplier for lung cancer that had been proposed by Dr. Selikoff . . . Walker allowed himself to be persuaded that Dr. Selikoff and his colleages at Mount Sinai's Environmental Sciences Laboratory had vastly overdiagnosed cases of mesothelioma. [16, July 1, 1985, pp. 45–6]

Here is pure ectoplastistics in its early stages of development. We have this unusual opportunity to witness the process because of the records of extensive legal proceedings before judges and juries. Dr. Walker's own sworn testimony gives a clear description of the fabrication of ectoplastistics:

> [Question to Walker:] Were you requested at any time by Manville or its counsel to make assumptions in your estimates that would result in lower rather than higher resulting numbers?
>
> Walker: I was asked . . . that whenever I had the chance to choose between two equally plausible assumptions, I should choose the assumption which led to the smaller number of cases of disease. [16, July 1, 1985, p. 46]

Thus, the numbers Johns-Manville presented to the court were pure ectoplastistics. More "certainty" meant taking the most favorable values from a range of values produced as the results of the most favorable assumptions.

What can and should be done in cases like this? The appropriate action is what epidemiologist Dreyer did in her original report: Give the upper and lower values produced by the estimates, describe in detail how they were obtained, and report their uncertainties. Then the readers (and if need be, judge or jury) can make their own judgments as to what to believe and conclude. This is what finally happened in the asbestos suits; judges and juries obtained enough information about the ectoplastistical processes to make decisions. But they have that information only because of the disclosures which were forced by the plaintiffs' attorneys, a tribute to both the legal system and the attorneys' diligence. We all need to exercise similar diligence in "busting" ectoplastistic ghosts.

VII. Guard Against the Changing of Numbers: David Stockman, Congress, and the Federal Budget Deficit

Do you like the numbers you have? If not, simply change them and emanate some ectoplastistics. This approach to misusing statistics seems to be more useful to agencies which produce statistics than to those who analyze the information. After all, if an analyst reports that he or she has examined a particular set of published statistics and tells how the analysis was done, then the calculations and interpretations can be checked by others. But if a public or private agency produces statistics—such as the numbers of employed or telephone calls made—it is virtually impossible for anyone else to verify those statistics. It is true that some checks are possible in some cases to determine whether or not the published statistics are reasonably correct. But without access to the sources of data and the process of collecting those data, there is no substantive way to check them.

Computer data processing and storage has made it easier to change numbers without detection. Data are stored in the form of magnetized spots on tapes or disks. If paper copies or special reports are not available, revisions of the magnetically stored data cannot be detected. Occasional unauthorized access to computer files by high school students make headlines, but these are infrequent occurrences and of small consequence compared to deliberate alteration of computer files and programs by people who have legitimate access to, or control over, the files, programs, and computer systems. As Robert H. Courtney, Jr., a former IBM security consultant said: "If you prioritize the security problem, they [computer whiz kids] might come right after leaky roofs and overflowing lavatories" [17, p. D-2].

There are few constraints to limit tampering with data or programs by people with access to them. Before computers, statistics scheduled for publication were typed and kept as documents ("hard copy" in computer jargon). If someone wanted to alter the figures or the computational formulas (such as are now carried out by computer programs) typists, clerks, mathematicians, and statisticians would participate in the change, and even if they didn't, there was usually some physical evidence of change (erasures, for example). Today, the magnetically stored data can be altered by those in the know with no record of the change.

David A. Stockman, the Reagan administration's first director of the Federal Budget Office, showed how to ectoplastistically get results desired by the administration. The emanatory process has been described in several places since Stockman's departure from office, but we draw on the contemporary account by William Geider [18, pp. 27 ff.].

In early 1981, President Reagan called for a three-year tax reduction and an increase in defense spending. To determine the effect of these significant actions on the country's economy, Stockman's staff fed the appropriate inputs and statistics into a computer model which was programmed to simulate the workings of the national economy. The results of this simulation predicted that the Federal budget deficits would be about $82 billion in 1982 and $116 billion in 1984. The opinion of the administration's financial advisors was that if such possible budgetary deficits were included in the President's first budget messsage in 1981, the country's financial market would panic.

Hence Stockman felt that he had to have projections that would show much smaller deficits. His staff modified the computer program to project much smaller budget deficits, a case of pure ectoplastistics. (A shame, since the original computer model was a good predictor.) The President's proposals for tax reduction and defense spending were accepted and the federal budget deficit in 1985 approached $200 billion.

But Congress was equally guilty of ectoplastistic emanation. Pressure to reduce the deficit continued, with Stockman leading the campaign with Congress. And in 1981, Congress authorized budget reductions of about $35 billion. Alas, pure ectoplastistics, for "The total of $35 billion was less than it seemed, because the 'cuts' were from an imaginary number—hypothetical projections from the Congressional Budget Office (CBO) on where spending would go if nothing changed in policy or economic activity" [18, p. 51]. A nice twist in the process of generating ectoplastistics. The reduction was from a nonexistent base. As

> Stockman explained: "There was less there than met the eye. Nobody has figured it out yet. Let's say that you and I walked outside and I waved a wand and said, I've just lowered the temperature from 110 to 78. Would you believe me? What this was was a cut from an artificial CBO base. That's why it looked so big. It wasn't." [18, p. 51]

Once again, the emanator of the ectoplastistics revealed the process of how to make and use ephemeral numbers. Unfortunately, the people who believed that the budget was actually reduced by $35 billion must be vast in number. Perhaps, sometime in the future, Stockman's frankness about the ectoplastistic process will make people more critical.

VIII. Summary

To determine whether a statistical statement is a spiritual emanation, examine the process by which it was generated. Always ask: *How do you know it?*

Some of the major issues for the reader are:

Estimates. How were they obtained? What arithmetic operations were performed? What are the corrections that were made to the input data? Who made them? Were the estimates the result of an agreement or in response to pressure?

Sources. Who made the estimates? What were the political or economic pressures on those who made the estimates? What is the established track record of those who made the estimates?

14
Big Brother/Sister: The Big Misuser

Every kind of government seems to be afflicted by some evil inherent in its nature.
Alexis de Tocqueville

Practical politics consists in ignoring facts.
Henry Adams

I. Introduction

The Inca civilization predated the Spanish empire in the Andes of Latin America. We have no evidence of writing *per se* in the archaeological finds for this civilization, but we have evidence that they kept statistical records. Their numbers were recorded on *quipus*, or knotted strings. We know that their records included, at a minimum, births and deaths, amount of food produced, amount and kinds of tribute brought to the ruler, and so forth [1, pp. 331–333].

The use of statistics by governments is as old as the formation of organized societies. Thousands of clay tablets found in the "cradle of civilization" (the region of the Tigris and Euphrates rivers) recorded the accumulations and disbursements of foodstuffs and other goods. Animal flocks, laborers' efforts, movements to and from stores, and other activities we now consider commercial but that were then under the control of government and the religious hierarchy, are the substance of the myriad tablets with numerical inscriptions carved on them. In the field of external relations, the records of military and acquisitive ventures, counts of prisoners taken, reports of sizes of armies, deaths, and areas lost or gained fill many stelae, panels, and graven records.

Over 3000 years ago, governments collected statistics to determine how many men could be called up for military service and how many taxpayers could be separated from a part of their earnings and wealth, and to prove that their countries were becoming overcrowded and must be expanded at someone else's expense [2, p. 360 ff.]. These form the substance of many of the records we now find of artistic as well as historical interest.

All of this is typical of the information any government needs to maintain control, administer an empire, and allocate resources to the governed; it is the raw material of governmental bureaucracy. But it is also the raw material of that catalogue of misuses of statistics by which a government can maintain its power and move and direct the governed—while also justifying the reward or punishment of groups, acquiring wealth by extracting it from the governed, and rationalizing internal repression and military and expansionary adventures.

We see these principles clearly displayed in the Inca example. The Inca government had complete control over the governed. The Inca Empire was

> authoritarian, bureaucratic and socialistic [in a way] that has not been approached by any state at any other time or place. . . . The Inca imperial government dictated to its subjects, in detail, the locality in which they were to live, the kind of work they were to do there, and the use that was to be made of the product of their labor. [1, p. xi]

While rulers may choose to use and misuse statistics as they wish, a government must have the straight facts for its internal use if it wants to function efficiently and effectively. We have interesting evidence of the efforts the Incas made to assure that their statistics were correct, especially at the local level where data were created:

> However small, it [the village] had at least four *quipucamayus* [the local statisticians who kept the records]. They all kept the same records and although one accountant or scribe was all that would have been necessary to keep them the Incas preferred to have plenty in each village and for each sort of calculation, so as to avoid faults that might occur if there were few, saying that if there were a number of them, they would either all be at fault, or none of them. [1, p. 331]

That the central government insisted on strict control measures may be our clue that local cheating existed, since a local authority whose power is less absolute might use statistical cheating to improve his position with respect to the central government or to accumulate private wealth. The use and misuse of statistics by governments is an ancient art.

The dominant constant in the government equation is that there is a government and there are those who are governed. The spectrum of governmental control ranges from governments that hold their power only by the common consent of the governed to governments that impose their absolute will without regard for either the opinions or lives of the governed. History gives many examples of these extremes, but most governments fall somewhere in between. A measure of support of the government by the governed is important for maintaining power. This is true even though cultural differences and political bias may blind observers in one society to the influence of the governed on the governors in another society.

It is because governments must convince some substantial segment of the governed that certain policies are ''right'' that governments turn to statistics, and, unfortunately, to misusing statistics. No one can safely say of misuses of statistics by governments, ''let me count the ways,'' because governmental misuses spread throughout all human history and are as numerous as the grains of sand on a beach. Despite our arrays of data banks, forests of publications, and armies of tabulators, future historians will have as much trouble figuring out precisely what happened, statistically, in the 20th century as we have deciding precisely what happened in the Inca empire.

We see three major schools of thought at work in governmental misuses of statistics which we call the *ectoplastistic*, the *closet*, and the *power-drunk*. In the ectoplastistic school, the government flaunts statistics, produces them continually in great quantity, and scatters them freely in all directions. If real numbers cannot be obtained, ectoplastistics will do:

> Capital Hill feeds on numbers, some authoritative, some spurious to buttress arguments and gain votes. . . . many are simply manufactured and convey ''an aura of spurious exactitude,'' in the words of Alice M. Rivlin, former director of the Congressional budget office, . . . the power of numbers in political debates has been noted by bureaucrats and lobbyists, members of Congress and Presidents, all of whom struggle to come up with the best numbers to make their cases. . . . ''A person who has control of numbers, whether accurate or not, will carry the day,'' said Herbert Kaufman, a political scientist. [3]

For a fuller discussion of governmental ectoplastistics, see Chapter 13.

The members of the closet school classify numbers as ''top secret'' and sequester them where the public cannot get at them. *They* will decide what numbers the general public is entitled to see. Even in the United States, such ''closeting'' actions occur again and again, as we know from the frequent challenges which concerned members of the public make. When one of our students sought information on public works expenditures from a city public works department, the department's representatives refused to give him permission to inspect the *public* document in which the figures had been published. The same document was available across the street in the public library!

The third school, the power-drunk, publishes and distributes statistics which it misinterprets deliberately or inadvertently in order to prove that it (the government) is in the right and should remain in power. In this chapter, we examine examples of each school.

II. Popular Information and Ways of Acquiring It: The Practices of Governments

A. Aid and Comfort to the Enemy, or Hindrance and Discomfort to Friends?

In 1753, members of the House of Lords refused to permit a census of the British population. They argued that open knowledge of the size of the Britain's population would reveal to potential enemies how small an army the British could muster [4]. For this reason (and probably for some others as well), the British population remained uncounted until 1801. Consequently, we have only "guesstimates" for the population of England and Wales in the middle of the 18th Century, when England's industrial revolution was in its early stages [5]. Because of this lack of sound statistics, it is difficult to gauge precisely the impact of the beginning of the industrial revolution on birth and death rates, on migration to and from England, and on social and economic conditions. For example, how much of the migration from England to the American colonies was due to the effects of the beginning of the industrial revolution?

We must ask: Would this census information really have given aid and comfort to the enemy? Or would it just have hampered the opposition? Or frustrated future historians? Was this fear justified?

Two hundred years later in April 1944, the House of Lords repeated the same argument in a discussion pertaining to the publication of industry statistics. "We cannot relax in any way our precautions in this or other matters . . . ," said Lord Templemore. Such statistics would, once again, give aid and comfort to England's enemies [4].

There is nothing new under the sun.

B. The Iron Census Curtain

No one government has a monopoly on the selective withholding of statistics—but some governments withhold more than others. The withholders may state explicitly that these statistics are vital to national security, or the reasons may be unstated and we must impute motives to the government. There are times when secrecy in the conduct of government is justified, but in any particular case we must ask: would publication really compromise national security or are the statistics withheld because they reveal a lack of good information? Or are they withheld so that no one will ever know whether they are good or bad?

In 1926 the USSR published quantities of population census data, but no comparable amounts have been seen since. Statistics from the 1959 and 1970 population censuses were published, but far less than in 1926. Even fewer have been published from the more recent 1979 census. Conspicuous for its absence is any information on the ages of the population.

Statistics other than those from the population censuses are equally sparse. Most published data are keyed to the Soviet Republics (governmental units roughly equivalent to states in the United States)—data such as numbers of births, deaths, energy consumption, and so forth. Until 1974 this information was also available for oblasts (divisions of the Republics roughly comparable to counties in the United States). After 1974 no information for oblasts was published.

There are only 16 republics in the USSR and one of them, the Russian

Soviet Federated Republic (which includes Moscow), contains over half the 262 million people in the country. Indeed, Moscow, with 8.5 million people, is larger than many republics, such as Estonia, which has only about 1.5 million inhabitants. Each republic is so diverse that very little fruitful study can be conducted on the basis of the republics alone. Information for smaller and more homogeneous areas is needed, and such information can only be obtained from oblasts or even smaller units.

Some examples of withheld information can be seen by comparing the questions asked of the people in the 1979 population census and the data published in the slim census report. Statistics on the age of the people were collected but never published; data for oblasts were collected and never published. Published statistics on the numbers of children born (data needed for a comprehensive study of the birthrate) and on the educational level of the population were so sparse that almost no analyses are possible. And so on and on.

Another clue to the availability of these unpublished statistics is found in articles and talks by some Soviet population experts. Viktor I. Perevedentsev, in an interview with Seth Mydans of the *New York Times* [6], spoke on such topics as the education of husbands and wives, the increase in one-parent families, the extent of job changes, village versus city population growth, and other topics. In the article "Social Revolution Sweeps Through Soviet Union," Perevedentsev mentions these and other findings, based on statistics to which he must have had access, but which were never made available to the outside world [6].

Robert A. Lewis [7] and other students of the USSR have postulated that the demographic patterns of change in the USSR may be quite similar to those experienced by the United States and western Europe during the 19th and early part of the 20th centuries. It would be interesting and informative to compare the USSR with these and other countries at different times to see how similar or different these "demographic and social revolutions" may be. But no reputable statistician dares make these comparisons, since the Soviets have published so few of the essential statistics.

We must also note that any statistics which the Soviet government thinks may reflect adversely upon it are rarely, if ever, published. For example, very few crime or accident statistics are published. When the infant death rate apparently increased in the early 1970s, the Soviets stopped publishing such information [8]. Actually, as Fred Grupp and Ellen Jones point out, the apparent increase was due to faulty statistics and not to a true increase in the death rate. The Kremlin should not have panicked.

The USSR does give some statistics to the United Nations, which prints them in its statistical publications. No analysis is performed by the United Nations and not enough detailed information is made available by the USSR to evaluate and interpret the totals, as we have been able to do to some of the U.S. statistics. What the internal and external public knows about the USSR is what the government chooses to let them know. Beyond that, it is all pure ectoplastistics, which includes the emanations of people on both sides of the political fence. The cause of international understanding is not advanced in this situation. Furthermore, it is not clear to us that the USSR has gained in security by withholding information and feeding the growth of ectoplastistics, which can be used against it as well as for it.

C. The United Nations Gets into the Act

The United Nations and its organizations (such as the International Labor, Food and Agricultural, and World Health Organizations) accept statistics as furnished by their member countries and then publishes them. In most cases, the United Nations treats these submissions as unquestioned statistical truths. But even the United Nations and its member organizations withhold statistics.

A recent illustration of this approach is the way in which the World Health Organization (WHO) reported the cholera epidemic in Africa in 1985. Cholera is a deadly disease if not promptly treated. A country is supposed to report its presence to WHO. The purpose of such reporting is to alert the world community that a highly communicable disease is epidemic in some region. But unless the country reports it, WHO does not *officially* know of its existence and does not report the presence of cholera in that country. Clifford May reports that Ethiopia experienced a disease, which they call "acute diarrhea," and that the Sudan experienced a similar disease, which they call "severe gastroenteritis" [9]. Western physicians say they have no doubt that these two disease reports refer to cholera:

> Officials of the World Health Organization claim to know nothing of cholera in Ethiopia and the Sudan, reflecting a provision in the organization's charter: Unless a government formally reports an outbreak, W.H.O. does not acknowledge its existence. [9]

This is one of the ways in which the United Nations and its member organizations misuse statistics.

The United Nations does publish considerable quantities of inferior-quality information. If there are gross and obvious problems with a nation's data, the United Nations may make suggestions to that nation, but it does not publish corrected data without the consent of that country. United Nations documents do sometimes carry notes on the quality of the statistics for a given country. Whether the United Nations is responsible for publishing inadequate data or for withholding statistics is beside the point: It does both. We reproduce here portions of a correspondence with the United Nations which clearly show how the United Nations acted to withhold statistics for Taiwan, thereby becoming an ally of one side in a political struggle.

Until 1972, Taiwan was treated as a separate country, and the United Nations and its member organizations published quantities of Taiwanese statistics. In 1972, the United Nations agreed to the demand of the People's Republic of China that Taiwan be declared part of the People's Republic and not a separate country, and stopped publishing statistics for that country.* Presumably, information about the millions of people in Taiwan is included in the overall (and unfortunately, meager) data for the People's Republic of China. We can only wonder how the People's Republic of China gets its statistical information about Taiwan, since they have no presence or cooperative arrangement of any kind with the government or peoples of Taiwan.

From correspondence with the United Nations in early 1975, we have the following letters:

> [To Secretary General Waldheim from A.J. Jaffe] For a number of years the U.N. Statistical and Demographic Year Books, the National Accounts Year Book and . . . other statistical publications, carried Taiwan as a separate geographic area. All manner of statistics about this were published, statistics which were very useful to those of us who are interested in international statistical comparisons. . . . Beginning in 1972 the United Nations stopped publishing separate data. . . . why did the United Nations stop publishing data for Taiwan? To say that Taiwan is part of [the People's Republic of] China is begging the issues since the United Nations does publish data for many areas which, politically, are parts or dependencies or territories of other countries. . . .
>
> [To A.J. Jaffe from S.A. Goldberg, Director, U.N. Statistical Office] You may recall that since the adoption of General Assembly resolution 2758 (XXVI) of 25 October 1971 . . . the United Nations considers Taiwan as a province of the People's Republic of China. Because of Taiwan's changed status, the treatment of its data was modified in accordance with the United Nations general publication policy. . . . Normally the United Nations publishes statistics for countries as a whole. In some circumstances, the United Nations publishes data for certain geographical regions of a country unless the country in question specifically requests that the global figures only should be published. In the case of the People's Republic of China, it initially objected strongly to any separate mention of Taiwan. However, in order to provide clarification in regard to the areas covered under the heading of China for statistical purposes, the Secretariat was able in 1973 to arrive at an understanding with the Government that, in United Nations statistical publications, data and estimates covering the People's Republic of China generally include Taiwan Province in the fields of statistics relating to population, area, natural resources, natural condtions such as climate, and so forth. In other fields of statistics, they do not include Taiwan Province unless otherwise stated.
>
> [To S.A. Goldberg, Director, U.N. Statistical Office from A.J. Jaffe] Your basic explanation [as to why data for Taiwan are omitted] is that the U.N. General

---

*Whether or not this is a reasonable *political* action is not our concern.

> Assembly declared that Taiwan no longer exists as a separate country and therefore you do not publish statistics for it. This, to me, is pure politicization of statistics. Political rather than technical considerations govern the nature of the statistics collected and their dissemination. . . . For years statisticians and social scientists have been using data re Taiwan for studying questions on fertility, economic development, education, mobility, and a whole host of other subjects. Previous U.N. publications were invaluable for this purpose. Now, for political reasons, such data are "non-existent" as far as the outside world is concerned. From the viewpoint of the statistician and student, such political designation of "non-existence" raises a fundamental question: to what extent may other political considerations which are much less visible than in the case of Taiwan, be affecting statistics which the U.N. publishes? Can we believe them and use them, or should we view them with suspicion?
>
> In short, I feel that politics rather than sound statistical-technical considerations govern the statistics which the U.N. publishes. [10]

D. How Much Secrecy Is Too Much?

> *A popular Government without popular information, or the means of acquiring it, is but a Prologue to a Farce or a Tragedy; or perhaps both.*
>
> *James Madison**

Suppression of, and the failure to publish, statistics about the United States is a misuse of statistics, as well as a danger to the country, as James Madison noted. The United States publishes—or otherwise makes available—many more statistics than the USSR. Nevertheless, the U.S. government still withholds and avoids acquiring some information, and as of the mid-1980s, is planning to withhold and avoid acquiring even more.† The control techniques are not as heavy-handed as those of the USSR, and we have many ways of detecting governmental efforts to reduce "popular information, or the means of acquiring it."

How do we know that data are being withheld or that attempts are being made to do so? First, check if the customary statistics are no longer forthcoming, whether they come out later than they used to in the past, and whether they are drastically curtailed. It is important to have a good memory if you want to avoid being the victim of governmental suppression and misuse of statistics.

Second, if your memory is not that good, look at the budget request. If the executive office of government asks for less money for the collection and

---

*Fourth President of the United States, as quoted in the Annual Report of the U.S. Government Printing Office, Fiscal Year 1984, p. 4.

†Public opinion in the United States often opposes the withholding of information. At this time there are many signs that the open public debate over the trend which we discuss is leading to a change in that trend.

distribution of statistics than in previous years, then it is probably trying to avoid acquiring and publishing some information. When the Congressional branch of the government agrees to a reduction in this funding, then we may be on the road "to a Farce or a Tragedy; or perhaps both," as James Madison emphasized.

The U.S. government withholds (or avoids acquiring) information via the budgeting process. It declares that it does not have the money to publish the information, or it argues that the people who want that information should collect it themselves, or it states that the government will collect and publish only a fraction of the data customarily collected and published. We cannot say for certain that it is the deliberate intention of the U.S. government to withhold information. We have no "smoking gun." But as shown below, we can see that by reducing the budget and eliminating statistical studies, the result is a *de facto* suppression of information.

If any government succeeds in reducing its published statistical information, then it can pontificate whatever it wishes: detecting ectoplastistics becomes that much harder. The government can make all types of political decisions and if challenged, can say "prove me wrong." Without the facts, it can be extremely difficult, if not impossible, to contest a particular political decision and "prove it wrong."

In constant 1980 dollars, the budgeted amount for major statistical agencies was reduced from $387 million to $305 million from 1980 to 1983. By 1985, it was modestly increased to about $320 million, but the executive branch requested only $305 million for 1986 [11, Table 1b, p. CRS-6]. How much money will be finally allocated we cannot say as we write this. We can only note that a serious attempt has been made to curtail the collection and dissemination of information.

In 1984, the statistical situation was described as follows:

> *General Trends*. The reductions in resources for statistical programs were most severe from FY1980* to FY1982 and budgets have recovered somewhat since then. In general, resource levels remain below what they were in FY1980. Not all resource reductions, however, indicate proportionate declines in data quality or quantity. In some cases, efficiencies in data collection or processing have been introduced to accommodate budget reductions. Developments in automation or other technical statistical applications have assisted considerably in specific programs. In other cases, programs have been cut back substantially: by eliminating entire programs; by reducing the frequency with which certain data are collected and produced; by reducing sample sizes and, thus, reliability; by reducing the scope of particular data collections; or by providing fewer analyses of data collected. [12, p. CRS-6]

---

*"Fiscal Year" 1980. Since 1976, the fiscal year of the Federal Government runs from Oct. 1 to Sept. 30.

At a Congressional committee meeting held March 16, 1982, Congressman Robert Garcia raised the following question in regard to the drastic cuts in the statistical budgets since 1980: "Why has the administration decided to cut its budget in this small corner of government? Do they really think it will save money? Or are they convinced that without the statistics the problems of the people will vanish?" [13, p. 2]. Cutting the statistics budget is a simple solution to some of a government's problems: no facts, no problems.

Since 1980, the collection of statistics has been cut back and many governmental publications of statistics have been eliminated. "It's a real trend of this Administration to limit public access to information," said Eileen D. Cook, [the associate executive director of the American Library Association]. . . . "The entire thrust is to reduce the accountability of Government." . . . Representative Thomas J. Downey said: "Restricting the flow of information in an open society is dangerous. It's antithetical to the free and open operation of Government" [14]. Ms. Cook further commented:

> The Reagan Administration's attempt to degrade and eliminate social and economic statistics collected for years by agencies of the Federal Government should come as no surprise. . . . Since [the Administration's] arrival there have been efforts to reduce the quality, scope and quantity of social and economic research. . . . A related attitude is the Administration's desire not to be bothered with facts. . . . As a consequence of the damage done since 1981, the reliability of conclusions drawn from analyses of our data has gone down. . . . This is just what the Administration desires. [15]

and

> What was first seen as an emerging trend in April 1981 has by December 1984 become a continuing pattern of the Federal Government to restrict Goverment publications and information dissemination activities. [16]

Unfortunately, the quality of statistics often is a measure of the budget provided by the politicians and of the extent to which the statisticians employed by those politicians (through the agency of the government) are allowed to act professionally. The governmental statisticians and other technicians who produce the statistics are, for the most part, competent and skilled professionals who can provide professional explanations and objective, considered interpretations of the numbers, if the politicians allow them to do so. But recently we have seen American politicians demanding political explanations and interpretations different from those of the professional statisticians (see the following section). Where conformance to such political demands was not given, we have witnessed reduced budgets, early retirements, and unexpected resignations.

III. Ingenious Interpretations and Other Alterations: The Power-Drunk

Alter: To make different without changing into something else; Castrate, spay; To become different.

Alteration: The result of altering.

Webster's Ninth New Collegiate Dictionary

What a government doesn't suppress, it can alter. And if a government cannot easily alter statistics, then it can always interpret them in such a way as to reinforce the government's position. We give a few examples in this section, limited only by space, and not by the number of examples available.

A. Close to Home: Some Recent Examples

1. *A Matter of Semantics?*

Prior to 1971, each time new labor statistics were released, the Commissioner of the Bureau of Labor Statistics (BLS) held press briefings at which the statistics were interpreted by professional statisticians. Of course, if their explanations were contrary to what the politicians wanted, there was conflict and open discussion in the public arena, but the politicians held the whip hand. In early 1972, the politicians discontinued the briefings. According to Lazare Teper,

> The discontinuance of BLS briefings came immediately after two episodes. . . . On February 5, 1971, in the course of the BLS press briefing, Assistant Commissioner Harold Goldstein described a 0.2 percentage point January decline in the unemployment rate as ''marginally significant'' while simultaneously the Secretary of Labor informed reporters at the White House that the drop had ''great significance.'' At the March 5, 1971 briefing Goldstein viewed another 0.2 percent decline in the unemployment rate, which was accompanied by a drop in employment and hours of work, as being ''sort of mixed,'' while simultaneously the Secretary of Labor described the situation as ''heartening'' and the Secretary of the Treasury as one that is ''rather pleasant.'' . . . But even suspension of press briefings did not eliminate possible conflict between the professional statisticians and the Administration. In its June 1971 press release, BLS explained that the reported decline in the seasonally adjusted unemployment rate may have been ''somewhat overstated because of the seasonal adjustment procedures and because more young workers than usual were in school during the survey week'' and that the updating of the seasonal adjustment factors will probably moderate the reported change, as indeed did occur. These valid technical comments apparently produced a reaction within the Administration. President Nixon was apparently furious and his views were conveyed privately to the Department of Labor. [17]

Was this no more than a semantic confusion over the difference between the meaning of the word ''significance'' to a statistician* and to a lay person? If

---

*As discussed in Chapter 12, Section II.B.

Commissioner Goldstein was referring to *statistical* significance, then he meant that, since the reported reduction of 0.2 percent was a sample statistic, it was quite unlikely that the number of U.S. unemployed had dropped. In that case, the Secretary of Labor had little basis for regarding this as a real drop, let alone a "greatly" significant drop.

In the unlikely event that Commissioner Goldstein (a professional statistician) was making a lay judgment of the significance of an estimated reduction of about 10,000 of the nearly 5,000,000 unemployed persons,* then many subjective factors enter into the discussion, such as whether you feel this was a "marginally significant" or greatly significant reduction.

Whether the basis for these conflicting interpretations was the difference between professional and lay interpretations of the word "significance," or between the Commissioner's and the Administration's subjective points of view, the Administration's actions speak for themselves: The briefings were discontinued.

Shortly after, Harold Goldstein left the Bureau of Labor Statistics, taking early retirement. Soon after, another statistician, Peter Henle, retired:

> Was Henle, who wrote an internal memorandum protesting the termination of press briefings, a victim of his "indiscretion"? . . . I do not know. . . . A number of other long term employees of the Bureau of Labor Statistics also left by either taking early retirement or by accepting jobs in the private sector or other agencies. Whatever caused Goldstein, Henle and others to leave BLS, their departures did not go unnoticed by the BLS staff or elsewhere in the Federal statistical establishment. To many, rightly or wrongly, it spelled out what might happen to those who inadvertently displease those in power while adhering strictly to professional standards of conduct. [17]

2. *Not Accidentally . . .*

On November 17, 1982, the then Secretary of Labor, Raymond Donovan, issued a release in which he claimed that the workplace injury rate for the first year of the Reagan administration (1981) had decreased significantly since the last year of the Carter administration: ". . . nearly every major indicator of on-the-job safety and health improved significantly. . . . during the first year of the Reagan Administration workers were less likely to be hurt" [19].

On the same day, the Department of Labor released a statistical report entitled "Occupational Injuries and Illnesses in 1981," prepared by professional statisticians and giving no one any credit for anything. This report simply revealed that the statistical injury rate fell a little more in the last year of the Carter administration than in the first year of the Reagan administration. Who should get more credit—Carter, Reagan, or neither? The year-to-year changes in

---

*The official estimate for 1971 of the number of unemployed persons was 4,993,000; and of employed persons, 86,929,000 [18, Tables 341 and 351].

the injury rate are about what one would expect from random fluctuations (see Chapter 10, Section II, Part B). What is more likely is that nothing happened between 1980 and 1981 and that neither president deserved credit or blame.

B. Different Lands, Same Customs

1. *Pakistani Economist Is Apt Pupil*

Washington is willing to share its skills in misusing statistics to make government look good with learners from the Third World. The journalist and writer Richard Reeves gives us an example when discussing Pakistani statistics [20, p. 39 ff.]. Pakistan's Minister for Planning and Development had worked for the World Bank, an agency of the UN, for 12 years and had been director of the Policy Planning and Program Review Department. This experience qualified him to design economic plans for the Pakistani government. When Reeves visited him, the Minister talked about Pakistan's great economic progress. The Minister saw enormous improvement when he compared conditions in 1970 (when he left for Washington) with those in 1982 (when he returned). To support his contentions, he gave Reeves a copy of Pakistan's Sixth Plan.

Reeves found that the statistics shown in the Sixth Plan did indeed show great improvement. And the reason? East Pakistan, now known as Bangladesh, was included in the earlier figures, but not included in the later ones. This region had been the poorest part of the combined country. When it separated from West Pakistan to become Bangladesh, it took its poverty, malnutrition, and underdevelopment with it. Yes, Pakistan has improved its condition—by surgical removal of the most impoverished part of the country. Whether conditions in the remaining parts of Pakistan are any different today than in 1970 is unknown to us.

Information from the United Nations Annual Yearbook supports Reeves' and our analysis of the reason for Pakistan's "improvement." In 1970, income per person in Pakistan was reported as about $175 (U.S. dollars) and in Bangladesh as about half that amount, $80.* In 1968, when East Pakistan was included with West Pakistan in the national accounts, reported income per person was $120 [21, International Table 1A for 1968 income; 22, Table 1 for 1970 income]. Hence, in two years—1968 to 1970—the income per person purportedly increased by 50%. This cannot be explained by any economic activity, including inflation. It can be explained only in terms of Bangladesh and its poor being included in the earlier computations and removed after the political breakaway.

---

*Although Bangladesh separated from West Pakistan in 1971, the economists estimated income per person for both West Pakistan and Bangladesh as of 1970. We could not find separate data for East Pakistan (now Bangladesh) prior to 1970.

2. Confucius Says . . .

Since before the days of Confucius and Mencius, the ancient Chinese philosophers, increases in population have been thought to indicate good government. The ruler who took good care of his subjects thereby increased their numbers, and increases in their numbers was *prima facie* evidence of good and proper government.

From the dawn of history until the 18th century, Chinese governments counted only the people they were interested in—those who could be taxed or impressed into service for the government.

> In 1712, however, the famous emperor Kang Hsi, of the Tsing Dynasty, after detecting the gross errors in the population reports, decreed that, thereafter, population figures should no longer be used as basis for the allotment of the poll tax and the land tax. This gave local officials greater freedom to boost population growth, especially in the provinces, in order to show the apparent material prosperity which was commonly associated with a large and growing population. From this time on, errors of another category appeared because of the tendency deliberately to exaggerate the numbers of people so as to please the reigning emperor. [23, p. 1,2]

The situation is not so different today—the last part of the 20th century. The purpose of governmental statistics is to please the rulers and keep them in power. In 1977 Leo A. Orleans [24] summarized available information and concluded that there were at least two forces working against the proper use of statistics in China. The first is traditional Chinese culture and thought which is indifferent to the idea of quantity—one number is as good as another. The second force is what might be called "government pride and face." The Chinese government "is extremely sensitive to anything that could be interpreted as a failure or even a weakness. Since statistics are the basic measure of success, their publication is closely controlled. Both as an aspect of security and as a manifestation of pride . . ." [24, p. 47].

By 1979, as was indicated by the *New York Times* headline, "Press in China admits to lies, boasts and puffery," Chinese newspapers were admitting that many of the success stories which they had printed (reporting, for example, huge fruit production increases) were boastful and untruthful, but that they were what the government wanted to hear [25].

Fox Butterfield, who spoke Chinese, interviewed members of the State Statistical Bureau of the People's Republic of China when he was a *New York Times* reporter in China, and supported this contention. He found the professional statisticians thought that: (1) many factories reported their planned targets rather than their actual production, as plans were often higher than actual production; and (2) agricultural statistics always showed increasing yields per acre as was desired by the government, and this was achieved by reporting less land in use than was actually farmed. The professional statisticians lacked enough personnel to check all reports they received, however [26].

The catalogue is long. Another *New York Times* reporter, C. S. Wren, found similar problems five years later. In 1984, the Chinese press admitted printing false statistics about the Dazhai commune which made it look like one of the most productive communes in the country [27]. A year later, Chinese officials claimed that the current governmental policies of economic liberalization have attracted eight billion dollars of foreign capital. But how is "capital" defined? Western embassies that track foreign investment in China estimate the amount of invested foreign capital is less than one billion dollars [28]. Do trade contracts and other financial arrangements—admittedly of great importance in China's international commercial relationships—account for the additional seven billion dollars? If so, they do not involve actual capital investment by foreigners in our sense of the term. Is this the consequence of legitimate differences in Chinese and western definitions of capital; or do they not deserve forgiveness, for they know what they do?

A last entry. The headline reads "China says economy soared in 1985." Note the careful wording [29]. China claimed that it had a very high growth rate. What is the story behind the headlines?

As nearly as the outside world can determine, Chinese governmental statistical practices have not changed in the 2500 years since the birth of Confucius. While the newspapers are taking the blame for the misuses of statistics in China, the power behind the headline is the government.

In the People's Republic of China, the government controls the newspapers. In the United States, the government does not control the newspapers; but it does control the budget of the statistical agencies—and behaves the same way as the Chinese and other governments do in trying to be the power behind the headline. It is the informed reader, like you, who can identify and analyze governmental misuses of statistics and who must try to keep government honest.

15
Concluding Remarks

You can fool all of the people some of the time, and some of the people all of the time, but you can't fool all of the people all of the time.
Abraham Lincoln

William F. Ogburn, president of the American Statistical Association in 1931, former editor of the *Journal of the American Statistical Association*, and a teacher at the University of Chicago, taught his students to ask three questions—to be used to determine the quality of an investigation and whether statistics were being used properly or not. The three questions were:

1. **What are you trying to find out**? What is the problem? Unless the aim of the work or question asked in the report or survey is clearly stated, forget any purported findings.
2. **How do you know it**? Once you have gotten the presumed answer, how do you know it? If statistics have been misused in the process of getting the answer, you do not necessarily have a correct answer and you probably have a wrong answer.
3. **What of it**? Once you have stated the problem to be solved or questions to be answered and obtained a solution or an answer, the question of importance and relevance must be answered. Was the problem important enough to be worth the effort?

In general, we have not discussed questions 1 and 3 in our book. These two questions are judgmental and often can be answered only in a particular context. Different people are interested in different subjects at different times, and both intent and importance are usually subjective. As long as the writer has clearly stated the problem which he or she seeks to answer, we have no complaints. How many ''dope addicts'' are there? How many people keep scorpions as pets? What is the balance of trade? How many high school dropouts are unemployed? Pick the question or problem which is relevant to you. We are only concerned that, if you or someone else uses statistics to get answers, the appropriate statistics are used correctly.

The entire contents of our book are concerned with asking and answering question 2: **How do you know it**? Our purpose in this book is to help you, the reader, judge whether appropriate statistics were used correctly. We highlighted the types of statistics of which you should be aware:

Surveys and Polls (Chapters 9 and 10)

Ockham's Razor (Chapter 11)

Affirmative Action and Discrimination Statistics (Chapter 12)

Ectoplastistics (Chapter 13)

Governmental Statistics (Chapter 14)

In the other chapters of this book we have briefly described—not too briefly, we hope—how the methods used for analyzing the basic data can be misused.

If you, the reader, insist on asking **How do you know it**? and demand the proof to back it up, you can be one of Abraham Lincoln's third category of ''some people'' who avoid being ''fooled all the time.'' That's a lot better for you (and for the country) than falling into Lincoln's first or second category.

And with this, we shave our dissertation with Ockham's Razor.

Appendix—Life Tables

Bad weather is preferable to no weather.

Eugene Michael Kulischer

I. Introduction

How long will you live? No one can tell you for sure. We only know how long the average person will live under the prevailing health and death conditions. That is what you can find out from a life table. Of course, no one is an average person; there is no "average person" who is certain to live exactly that time shown in the life table. After all, there are infant deaths and deaths of persons who live more than 100 years. Properly speaking, we can only find the average duration of a lifetime from the tables. However, since it simplifies the explanation, we will talk in terms of an average person. Knowing something about this hypothetical average person is more useful than no knowledge whatsoever.

In this appendix, we briefly explain the general concept of life tables. The construction of a life table is complicated, and we do not attempt to describe the process. The general concept is all you need to read and understand the most common life tables and values derived from them, such as life expectancy or length of life. Most frequently, you will want to know two values: (1) how long the average person will live from birth to death; and (2) how long the average person will live from some specified age (such as 20, or 60) to death. No special mathematical or statistical skills are needed to understand how to get and use this information.

Table A.1 A Hypothetical Cohort Life Table

Time period	Age	Size of cohort at beginning of period	Number of deaths in cohort during period
Jan. 1, 1900—Dec. 31, 1900	0	10,000	132
Jan. 1, 1901—Dec. 31, 1901	1	9868*	8
Jan. 1, 1902—Dec. 31, 1902	2	9860	7
Jan. 1, 1903—Dec. 31, 1903	3	9853	6
Jan. 1, 1904—Dec. 31, 1904	4	9847	4
Jan. 1, 1905—Dec. 31, 1905	5	9843	. . .
. . .	. . .	. . .	. . .
Jan. 1, 2000—Dec. 31, 2000	100	0	0
Total Deaths			10,000

*10,000 minus 132.

II. Everyone's in Step: The Cohort Life Table

Imagine that you are reading this on the first instant of the day January 1, 1900, and at this instant 10,000 babies are simultaneously born in the First General Hospital of Central City, USA. We call these 10,000 newborn infants a *cohort*. The number you choose to imagine can be 100,000, 1,000,000, or whatever number is conceptually convenient. The exact amount is irrelevant to the explanation. Now let us see what happens to our cohort between the starting date of January 1, 1900, and the time when the last cohort member dies. To show how cohort analysis works, assume the deaths shown in Table A.1 occur in our cohort. As you can see, the table continues until the last member of the original cohort has died sometime around the year 2000. There is a total of 10,000 deaths, equal to the 10,000 births that formed the cohort at its start.

III. The Current Life Table

Most people are more interested in the present than in the past. The death rate has decreased considerably since 1900. Consequently, we want a life table that will tell us what is happening *now* in the latter quarter of the 20th century. Actuaries devised the current table for this purpose. Hence, most life tables that the reader is likely to see are "current" tables rather than cohort tables. The basic idea is exactly the same as we just described for cohort life tables, differing only in some details of the mechanics of construction, which we now describe.

To calculate a current life table, the actuary begins with the population and numbers of reported deaths (or death rates) for each age as of some specified date.* The baseline statistics might be for a date (such as the year 1986) or based on a three-year average (i.e., 1985 to 1987). The calculations can be made for the entire population of men and women, for men only, for women only, or for any other part of the total population which is of interest and for which information on deaths is available. From this point on the logic is exactly the same as for a cohort life table.

IV. Length of Life

A. From Ground Zero

How do we determine how long *you* (the average person of the cohort) might live—your life expectancy or length of life—from the numbers in the table? It's simple! In 1901 there were 9868 babies celebrating their first year of life; this is a *credit* of 9868 *person-years*. What about the 132 who didn't make it through the first year? We assume an average lifetime for each of these 132 of one-half year,† which gives an additional credit of 66 (132 times ½) person-years; the total is 9934 person-years (9868 plus 66).

In 1902, there were 9860 babies who lived to their second year, contributing 9860 person-years to the total. We assume that eight died in that year; they contribute four person-years to make the 1902 credit equal to 9864 person-years. Thus, at the end of 1902, the total person-years accumulated is 19,798 (9934 plus 9864).

Continue these calculations until all members of the cohort of 10,000 babies have died, and you will have the total number of person-years lived by the members of the cohort. Let us assume that in this case the total is 740,000 person-years. Then the average member of the cohort has lived 740,000 divided by 10,000 (the number of births), or 74 years. This figure generally appears in the right-hand (or last) column of a published life table.

B. Starting from Now, or Any Age

These calculations can start at any age. If you start with age 20, for example, you may find that life expectancy is, on the average, 56 years. This means that the (nonexistent) average 20-year old person can expect to live to 76 (20 plus 56).

---

*These data are available from the U.S. National Center for Health Statistics.

† One-half (½) is an approximation which we use here for simplicity in order to explain the concepts. Actuaries use more precise fractions for the first few years of life. For older ages, actuaries usually use the fraction one-half year.

V. Life Table Rates: Birth and Death

The life table is constructed such that its population never changes in size. Hence, if a population is to stay constant, then the deaths which we describe in the table above must be balanced each year by a number of births equal to the number of deaths. The birth and death rates as calculated from a life table must be equal. What is the average rate of births per year that will balance the deaths? It is simply the reciprocal of the life expectancy. If the life expectancy is 74 years, then the average birthrate and death rate must be .0135 (1 divided by 74) per year, or 13.5 per 1000 population [1, Table 104].

Note that these life table rates are not necessarily the same as those reported for the actual population. For example, in the United States in 1981, the life table birthrate was 13.5 per 1000 population, whereas the actual birthrate was 15.8 [1, Table 80]. The life table death rate was the same as the birthrate, 13.5, whereas the actual death rate was 8.6 per 1000 population. Thus, the actual population of the United States was increasing by 7.2 persons per 1000 population per year, whereas the life table population neither increases nor decreases.

VI. In Conclusion

There are many additional aspects of the interpretation and analysis of real life tables which we do not discuss here. However, it is interesting to note that the same principles we discussed here in regard to people apply to (and are used in) analyzing and predicting the life of manufactured equipment and goods. The next time you buy an electric light bulb, look on the box for the manufacturer's statement of the average lifetime for that type of bulb.

References

CHAPTER 1: INTRODUCTION

1. Hershey H. Friedman and Linda W. Friedman, A new approach to teaching statistics: learning from misuses, *New York Statistician, 31*(4–5):1 (1980).
2. I. Bernard Cohen, Florence Nightingale, *Scientific American, 250*(3):128 (1984).
3. Stephen Jay Gould, Singapore's patrimony (and matrimony), *Natural History, 93*(5):22 (1984).
4. Herbert F. Spirer and A. J. Jaffe, *New York Statistician, 32*(3):1 (1981).
5. Stephen Jay Gould, Morton's ranking of races by cranial capacity, *Science, 200*(4341):503 (1978).
6. Leonard Silk, Economic scene. Call to hire 'One Person,' *New York Times,* December 29, 1982.
7. Louis Pollack and H. Weiss, Communication satellites: countdown for Intelsat VI, *Science, 223*(4636):553 (1984).
8. Letters to the editor, *Science, 224*(4648):446 (1984).

CHAPTER 2: CATEGORIES OF MISUSE

1. G. Hardin, No need for a census, use samples, *New York Times,* Feb. 7, 1980.
2. Phony formulas, *Wall Street Journal,* April 2, 1979.
3. Ronald C. Moe, The empty voting booth: fact or fiction?, *Commonsense, 2*(1):11 (1979).
4. Center for Science in the Public Interest, National Nutrition Referendum, *Nutrition Action.*
5. E. J. Dionne, Jr., Abortion poll: not clearcut, *New York Times,* Aug. 18, 1980.
6. Henry Scott Stokes, Japan's trade figures: a matter of accounting, *New York Times,* July 22, 1979.

7. Loy I. Julius, Richard W. Hungerford, William J. Nelson, Theodore McKercher, and Robert W. Zellhoefer, Prevention of dry socket with local application of Terra-Cortril in Gelfoam, *Journal of American Oral and Maxillofacial Surgeons, 40*:285 (1982).
8. Herbert F. Spirer, S. Rappaport, and A. J. Jaffe, Misuses of statistics: the jaws of man I, *New York Statistician, 34*(5):3 (1983).
9. Minimum-competency tests having minimum effects, *New York Times,* April 22, 1979.
10. Why isn't America using more coal?, newspaper advertisement; source of data, probability sample by Roger Seasonwein Associates, Inc., May 1979.
11. Malcolm W. Browne, When numbers just don't add up, *New York Times,* April 22, 1980.
12. Jerry E. Bishop, Surgeons find heart repair pays for itself, *Wall Street Journal,* Aug. 29, 1980.
13. Glenn Collins, Migration trend of the aging, *New York Times,* Oct. 2, 1984.
14. Karen W. Arenson, Martin Feldstein's computer error, *New York Times,* Oct. 5, 1980.
15. Sam Rosensohn, Happiness is . . . being able to say 'no' to sex, *New York Post,* Sept. 16, 1980.
16. Joyce Purnick, Convictions of 3.8% found in updated day-care check, *New York Times,* Feb. 18, 1985.
17. Andy Logan, Around City Hall: challenges, *New Yorker,* Aug. 19, 1985, p. 78.
18. Advertisement, *New York Times Magazine,* July 15, 1979.
19. Shanna H. Swan and Willard I. Brown, Vasectomy and cancer of the cervix, *New England Medical Journal, 301*:46 (1979).
20. Phillip Shabecoff, Jobless rate in U.S. up slightly from July, to 5.7% from 5.6%, *New York Times,* Aug. 4, 1979.
21. Kathleen Stein, Dr. C's vitamin elixers, *OMNI,* (issue unknown), p. 69 (1982).
22. Gina Kolata, Computing in the language of science, *Science, 224*(4645):150 (1984).

CHAPTER 3: KNOW THE SUBJECT MATTER

1. U.S. Bureau of the Census, *1970 Census,* Vol. PC(2)-1F, p. X.
2. League of Nations, *International Statistical Year-book,* 1928.
3. United Nations, *Demographic Year Book,* 1948.
4. United Nations, *Demographic Year Book,* 1964.
5. Conrad Taeuber and Irene B. Taeuber, *The Changing Populations of the United States,* Wiley, New York, 1958.
6. Robert A. Lewis and Richard H. Rowland, *Population Redistribution in the USSR,* Praeger, New York, 1979.
7. Donald D. Atlas, Music's charms may lengthen life, *New York Times,* Dec. 5, 1978.
8. J. Douglas Carroll, Letter to the editor, *New York Times,* Jan. 23, 1979.
9. U.S. Department of Health and Human Services, Public Health Service, National

Center for Health Statistics, *Vital Statistics of the United States: 1978,* Vol. II, Section 5.
10. Peter Hubbard, *Science, 217*(4563):919 (1982).
11. *New York Times,* Oct. 30, 1979.
12. U.S. Bureau of the Census, *1970 Census,* Vol. PC(2)-7A.
13. U.S. Office of Education, *Earned Degrees.* (See annual issues.)
14. F. A. Hassan, *Demographic Archaeology,* Academic Press, New York, 1981, p. 128.
15. U.S. Department of Health and Human Services, Public Health Service, National Center for Health Statistics, *Vital Statistics of the United States 1979,* Vol. II, Section 6.
16. U.S. Bureau of the Census, *Statistical Abstract of the United States: 1985.*
17. Kenneth M. Weiss, Demographic models for anthropology, *Memoirs of the Society for American Anthropology,* April 1973.
18. United Nations, *The Determinants and Consequences of Population Trends,* Vol. I, New York, 1973.
19. U.S. Bureau of the Census, Nation to reach zero population growth by 2050, Release, Nov. 9, 1982.
20. M. H. Morrison, *Economics of Aging,* Van Nostrand Reinhold, New York, 1982.
21. William Bradley, Letter to New Jerseyans, Feb. 1983.
22. U.S. Bureau of the Census, Prospectives of the Population of the United States: 1982 to 2050 (Advance report), *Population Estimates and Prospectives,* Series P-25, no. 922, Oct. 1982.
23. Barbara Rieman Herzog (Ed.), *Aging and Income,* Human Sciences Press, New York, 1978.

CHAPTER 4: DEFINITIONS

1. U.S. Bureau of the Census, *Statistical Abstract of the United States: 1985.*
2. U.S. Bureau of the Census, *1980 Census of Population, Characteristics of the Population, General Social and Economic Characteristics U.S. Summary,* Vol. 1, Chapter C, Part 1, PC 80-1-C1, 1983.
3. Constitution of the United States, Article I, Section II, Part 3.
4. Producers' prices show slight drop, first in 4½ Years, *New York Times,* Oct. 4, 1980.
5. Richard Corrigan, Aerospace industry waits for Bureau of Labor Statistics to fill in the blank, *National Journal* (Aug. 17, 1985), pp. 1888–1890.
6. United Nations, *Statistical Year Book,* New York, 1981.
7. United Nations, *Demographic Year Book,* New York, 1981.
8. Albert Shanker, How real is the chancellor's dropout report?, *New York Times,* Oct. 28, 1979.
9. Study disputes earlier questions on visits to prisons by youths to reduce Jersey crime, *New York Times,* Jan. 13, 1980.
10. Henry Scott Stokes, Japan's trade figures: a matter of accounting, *New York Times,* July 23, 1979.

11. Sandra R. Gregg, Mayor announces 30% drop in TB rate, *Washington Post,* Apr. 14, 1982.
12. David Krivine, How many emigrants are there really?, *Jerusalem Post* (International Edition), Sept. 14, 1985.

CHAPTER 5: THE QUALITY OF BASIC DATA

1. Max Singer, The vitality of mythical numbers, *The Public Interest, 23*:3 (1971).
2. A. J. Jaffe, Some observations on the nature of basic quantitative data, *New York Statistician, 33*(5):5 (1982).
3. A. J. Jaffe, Misuse of statistics XVIII—misusing statistics to solicit money, *New York Statistician, 33*(3):3 (1982).
4. U.S. Bureau of the Census, *Statistical Abstract of the United States: 1985.*
5. Herbert F. Spirer and A. J. Jaffe, Misuse of statistics XII, *New York Statistician, 32*(3):1 (1981).
6. Mark L. Trencher, Letter to the editor, *New York Statistician, 32*(4):7 (1981).
7. U.S. Bureau of the Census, *Historical Statistics of the U.S. Colonial Times to 1970,* 1975.
8. U.S. Bureau of the Census, *Statistical Abstract of the United States: 1939.*
9. U.S. Bureau of the Census, *Statistical Abstract of the United States: 1978.*
10. U.S. Bureau of the Census, *Statistical Abstract of the United States: 1979.*
11. John Head, 20 million illegal aliens get reprieve, *Denver Post,* Sept. 7, 1980.
12. Jeffrey S. Passel, ''Undocumented Immigrants: How Many?,'' 1985 Annual Meetings of the American Statistical Association, Las Vegas, 1985.
13. *Philadelphia Bulletin,* Oct. 21, 1979.
14. Melvin A. Benarde, Letter to the editor: food additives, *Science, 206*(4424):206 (1979).
15. Kathleen Stein, Dr. C's vitamin elixers, *OMNI,* (issue unknown) p. 69 (1982).
16. U.S. Public Health Service, *U.S. Life Tables, 1949–51, Vital Statistics Special Reports, 41*(1):8 (1954).
17. U.S. Bureau of the Census, *Statistical Abstract of the United States: 1984.*
18. Murray Feshbach, The age structure of the Soviet population: preliminary analysis of unpublished data, *Soviet Economy, 1*:178.
19. W. Ward Kingkade, *Evaluation of Selected Soviet Population Statistics,* U.S. Bureau of the Census, 1985.
20. U.S. Federal Bureau of Investigation, *Uniform Crime Reporting Handbook,* 1966.
21. Lawrence Sherman and Barry Glick, The quality of police arrest statistics, *PF Reports,* Aug. 1984, p. 1.
22. David Burnham, F.B.I. says 12,000 faulty reports on suspects are issued each day, *New York Times,* Aug. 25, 1985.
23. Phillip M. Boffey, Panel finds no fraud by alcohol researchers, *New York Times,* Sept. 11, 1984.
24. Paul R. Fish, Consistency in archeological measurement and classification: a pilot study, *American Antiquity, 43*:86 (1978).
25. U.S. Bureau of the Census, *1970 Census of Population, Characteristics of the*

Population, General Social and Economic Characteristics U.S. Summary, PC 70 (2)-1-F.

26. U.S. Bureau of the Census, *1980 Census of Population, Characteristics of the Population, General Social and Economic Characteristics U.S. Summary,* PC 80-1-C1.
27. U.S. Bureau of Indian Affairs, *Indian Service Population and Labor Force Estimates.*
28. Douglas H. Ubelaker, Estimates of Western Hemisphere population at the time of European discovery, *American Journal of Physical Anthropology, 45*(3):661 (1976).
29. Bureau of Applied Social Research of Columbia University, *Puerto Rican Population of New York City,* (A. J. Jaffe, ed.), New York, 1954.
30. Survey reports fertility levels plummet in developing nations, *New York Times,* Aug. 10, 1979.
31. Joseph A. Cavanaugh, Is fertility declining in less developed countries? An evaluation analysis of data sources and population programme assistance, *Population Studies, 33*(2):283 (1979).
32. Alfred L. Malabre, Thanks to off-the-books income, consumers save more than meets the eye, economists say, *Wall Street Journal,* June 11, 1982.
33. Denise B. Kandel, ed., *Longitudinal Research on Drug Use,* Wiley, New York, 1978.
34. John A. O'Donnell, Variables affecting drug use (book review), *Science, 203*(4382):739 (1979).
35. Denise Kandel, Stages in adolescent involvement in drug use, *Science, 190*:912 (1975).
36. Joseph A. Raelin, *Building a Career: The Effect of Initial Job Experiences and Related Work Attitudes on Later Employment,* Upjohn Institute for Employment Research, Kalamazoo, Mich., 1980.
37. Anonymous reviewer, private communication, Sept. 1985.
38. Sey Chassler, The Redbook report on sexual relationships, *Redbook Magazine,* Oct. 1980.
39. Roger S. Bagnall, For young classicists, a silver lining, *New York Times,* Sept. 18, 1985.

CHAPTER 6: GRAPHICS AND PRESENTATION

1. Gerardine DeSanctis, Computer graphics as decision aids: directions for research, *Decision Sciences, 15*:463 (1984).
2. Edward R. Tufte, *The Visual Display of Quantitative Information,* Graphics Press, Cheshire, Conn., 1983.
3. Howard Wainer, How to display data badly, *American Statistician, 38*(2):137 (1984).
4. Martin Mayer, Tremor in Orange County, *Barron's,* Jan. 16, 1978.
5. U.S. Department of Labor, Bureau of Labor Statistics, *Consumer price index for urban consumers (CPI-U), U.S. city average, All items, Series A,* 1979.
6. Where two aggressive companies plot growth, *Business Week,* June 16, 1980.

7. Frederick C. Klein, Americans hold increasing amounts in cash despite inflation and many other drawbacks, *Wall Street Journal,* July 5, 1979.
8. U.S. Bureau of the Census, *Statistical Abstract of the United States: 1985.*
9. David E. Moore, Market forces reshape health care industry, *Connecticut Business Journal,* Apr. 3, 1984.
10. *A.F.L.-C.I.O. News,* Feb. 3, 1979.
11. Steven J. Marcus, Solar-age windows, *New York Times,* Apr. 21, 1983.
12. U.S. Bureau of the Census, *Statistical Abstract of the United States: 1984.*
13. James Black, Misuse of statistics 27—representing a loss as a gain, *New York Statistician, 35*(4):3 (1984).

CHAPTER 7: METHODOLOGY

1. U.S. Bureau of the Census, *County and City Data Book, 1983.*
2. U.S. Bureau of the Census, *Statistical Abstract of the United States: 1985.*
3. American Society for Quality Control, *Quality Progress* (issue unknown).
4. Leonard Silk, Economic scene: the search for unity, *New York Times,* Sept. 24, 1980.
5. Karen W. Arenson, Economics: Martin Feldstein's computer error, *New York Times,* Oct. 5, 1980.
6. Leonard Silk, Economic scene: Social Security impact a puzzle, *New York Times,* Dec. 17, 1982.
7. Dean R. Leimer and Selig D. Lesnoy, What they found . . . , *New York Times,* Oct. 5, 1980.
8. Martin S. Feldstein, . . . And his defense, *New York Times,* Oct. 5, 1980.
9. U.S. Bureau of the Census, *Statistical Abstract of the United States: 1984.*
10. Frederick Mosteller and John W. Tukey, *Data Analysis and Regression,* Addison-Wesley, Reading, Mass., 1977.
11. Barbara Pitcher, *Summary report of validity studies carried out by ETS for graduate schools of business 1954–1970,* Educational Testing Service, Princeton, N.J., 1971.
12. Sandra Rosenhouse-Persson and George Sabagh, Attitudes toward abortion among Catholic Mexican-American women: the effects of religiosity and education, *Demography, 20*(1):87 (1983).
13. Mary Ann Chiasson and Stan Altan, Misuse of statistics, 30: smoking and face lifts, *New York Statistician, 36*(3):3 (1985).
14. Smoking and aging skin, *New York Times,* Sept. 25, 1984.
15. J. Neyman and E. S. Pearson, *Joint Statistical Papers,* University of California Press, Berkeley, 1967.

CHAPTER 8: FAULTY INTERPRETATION

1. Fujiya Hotel, Ltd., *We Japanese, Being Descriptions of Many of the Customs, Manners, Ceremonies, Festivals, Arts and Crafts of the Japanese,* Miyanoshita, Japan, 1950.
2. John D. Durand, Development of the labor force concept 1930–1940, *Labor Force Definition and Measurement,* Social Science Research Council, New York, 1947.

3. A. J. Jaffe and Charles D. Stewart, *Manpower Resources and Utilization,* Wiley, New York, 1951.
4. U.S. Department of Labor, *Employment and Earnings,* 1985.
5. U.S. Department of Labor, *Employment and Earnings,* 1983.
6. M. T. Kaufman, No longer a charity case, India fills its own granaries, *New York Times,* Aug. 10, 1980.
7. *World Development Report,* Oxford University Press, New York, 1983, Table 24.
8. U.S. Bureau of the Census, *Statistical Abstract of the United States: 1963.*
9. United Nations Food and Agriculture Organization, *1984 State of Food and Agriculture,* New York, 1984, Annex Tables 2 (food production) and 3 (total agricultural production).
10. Raymond Bonner, Gains from El Salvador land distribution disputed, *New York Times,* Apr. 19, 1982.
11. Edwin B. Williams (ed.), *The Scribner-Bantam English Dictionary,* Charles Scribner's Sons, New York, 1977.
12. Mark L. Trencher, Misuse of statistics 21, *New York Statistician, 34*(1): 5 (1982).
13. AFL-CIO Public Employee Department, *REVENEWS,* Feb. 1982.
14. Franklin L. Leonard, Misuse of statistics XIV, *New York Statistician, 32*(4):2 (1981).
15. Stephen R. Kellert and Alan R. Felthaus, Childhood cruelty toward animals among criminals and noncriminals, manuscript submitted to *Archives of General Psychiatry,* November 1983. Private communication.

CHAPTER 9: SURVEYS AND POLLS, PART I

1. U.S. Bureau of the Census, *1980 Census of Population, Number of Inhabitants U.S. Summary,* Vol. 1, Chapter A, PC 80-1-A1, 1983.
2. The man who knows how we think, *Modern Maturity,* April–May, 1974, p. 11.
3. Archdiocese of New York, Office of Pastoral Research, *Hispanics in New York: Religious, Cultural and Social Experiences,* Vol. 1, New York, 1982.
4. Norman Bowers and Francis W. Horvath, Keeping time: An analysis of errors in the measurement of unemployment duration, *Journal of Business and Economic Statistics, 2*:2 (1984).
5. Federation Employment and Guidance Service of New York, *Survey of Employers' Practices and Policies in the Hiring of Physically Impaired Workers,* A. J. Jaffe, (dir.), New York, 1959.
6. Robert Reinhold, Polls' divergence puzzles experts, *New York Times,* Aug. 15, 1984.
7. N. M. Bradburn, S. Sudman, and associates, *Improving Interview Method and Questionnaire Design,* Jossey-Bass, New York, 1980.
8. N. Schuman and S. Presser, *Questions and Answers in Attitude Surveys,* Academic Press, New York, 1981.
9. the *province,* Vancouver, British Columbia, Aug. 14, 1979.
10. E. J. Dionne, Jr., Abortion poll: not clear-cut, *New York Times,* Aug. 18, 1980.
11. Frederick Mosteller, Herbert Hyman, Phillip J. McCarthy, Eli S. Marks, and David

Truman, The pre-election polls of 1948, *Science Research Council Bull. 60,* New York, 1949.
12. J. Hachigian, Misuse of statistics XI, Carter-Reagan debate and the ABC telephone poll, *New York Statistician, 32*(2):1 (1980).
13. The *Playboy* reader's sex survey, *Playboy, 31*(1):108 (1983).
14. *Congressman Hallenbeck's Washington Report,* June, 1980, p. 4.
15. A. J. Jaffe, Not everyone has a telephone at home, *New York Statistician, 35*(5):2 (1984).
16. Roger L. Jenkins, Richard C. Reizenstein, and F. G. Rodgers, Report cards on the MBA, *Harvard Business Review, 62*(5):20 (1984).
17. U.S. Bureau of the Census, *Employment and Earnings,* 1983.
18. Century Opinion Polls, Inc., Report of Registered Democrats Regarding the 1980 Presidential Primary, *New York Poll,* May 25, 1979.
19. Research and Forecasts, Inc., *Aging in America: Trials and Triumphs,* New York, 1980.
20. Airline Passengers Association, Inc., *1979/1980 Membership Survey,* Dallas, Texas, 1980.

CHAPTER 10: SURVEYS AND POLLS, PART II

1. Lawrence Kilman, Poll: use rats, not dogs, for medical testing, *Stamford Advocate,* Oct. 29, 1985.
2. U.S. Bureau of the Census, Money Income of Households, Families, and Persons in the United States: 1983, *Consumer Income,* Series P-60, no. 146, 1985.
3. Jobs: unemployment is down, but not in Detroit, *New York Times,* December 9, 1979.
4. U.S. Bureau of Labor Statistics, *Geographic Profile of Employment and Unemployment: States, 1978; Metropolitan Areas, 1977–78,* 1978.
5. U.S. Bureau of Labor Statistics, *Employment and Earnings,* 1983.
6. Daniel Yankelovich, The polls don't mean a thing, *New York Times,* Oct. 7, 1979.
7. Abert H. Cantril, The polls shouldn't govern the debate, *New York Times,* Sept. 7, 1980.
8. Barbara A. Bailar and C. Michael Lanphier, *Development of Survey Methods to Assess Survey Practices,* American Statistical Association, Washington, D.C., 1978.
9. How the poll was conducted, *New York Times,* Aug. 14, 1984.
10. U.S. Bureau of the Census, *Statistical Abstract of the United States: 1984.*
11. A. J. Jaffe, Not everyone has a telephone at home, *New York Statistician, 35*(2):5 (1984).
12. A. J. Jaffe, Walter Adams, and Sandra G. Meyers, *Negro Higher Education in the 60s,* Praeger, New York, 1968.
13. A. J. Jaffe, Lincoln Day, and Walter Adams, *Disabled Workers in the Labor Market,* The Bedminster Press, Totowa, N.J., 1964.

CHAPTER 11: THE LAW OF PARSIMONY

1. John Kenneth Galbraith, in an article in the *Atlantic Monthly* as quoted in the *Wall Street Journal,* issue unknown.

2. Edi Karni and Barbara K. Shapiro, Tales of horror from ivory towers, *Journal of Political Economy,* Feb. 1980.
3. American Anthropological Association, Report of the ad hoc committee to implement the 1972 resolution on fair employment practices in employment of women, *Anthropology Newsletter, 20*:7 (1979).
4. A. J. Jaffe, Jaffe disagrees with committee's procedures, *Anthropology Newsletter, 20*(7):4 (1979).
5. David H. Thomas, The awful truth about statistics in archaeology, *American Antiquity, 43*(2):231 (1978).
6. Joel Gunn, *American Antiquity, 40*(3):21 (1975).
7. W. Brass and A. J. Coale, Methods of analysis and estimation, *The Demography of Tropical Africa,* W. Brass (ed.), Princeton University Press, Princeton, N.J., 1968.
8. W. Brass, Indirect methods of estimating mortality illustrated by application to Middle East and North Africa data, *The Population Framework, Demographic Analysis, Population and Development,* UNECWA, Beirut, Lebanon, 1978.
9. W. Brass, *Methods for Estimating Fertility and Mortality from Limited and Defective Data,* Laboratory for Population Studies, University of North Carolina, Chapel Hill, N.C., 1975.
10. Michel Garenne, Problems in applying the Brass method in tropical Africa: A case study in rural Senegal, *Genus,* Consiglio Nazionale delle Richerche (ed.), *XXXVIII*(1–2), Comitato Italiano per lo Studio dei Problemi della Popolazione, Rome, 1982, p. 119.
11. Wassily Leontief, Academic economics, Letter to the Editor, *Science, 217*(4555):104 (1982).
12. Richard A. Staley, Nonquantification in economics, Letter to the Editor, *Science, 217*(4566):1204 (1982).
13. Gerald A. Bodner, Point of view: should statistics alone be allowed to prove discrimination? *Chronicle of Higher Education,* June 22, 1983.
14. Wesley A. Fisher, Ethnic consciousness and intermarriage: Correlates of endogamy among the major Soviet nationalities, *Soviet Studies, XXIX*(3):395 (1976).
15. Brian D. Silver, Ethnic intermarriage and ethnic consciousness among Soviet nationalities, *Soviet Studies, XXX*(1):107 (1977).
16. Raymond P. Mayer and Robert A. Stowe, Would you believe 99.9969% explained?, *Industrial and Engineering Chemistry, 61*(5):42 (1969).

CHAPTER 12: AFFIRMATIVE ACTION AND DISCRIMINATION

1. U.S. Department of Labor, *Office of Federal Contract Compliance Programs ("O.F.C.C.P.") Rules and Regulations,* 41-CFR 60-1 (as amended).
2. U.S. Department of Labor, *Office Federal Contract Compliance Programs Affirmative Action Guidelines,* 41-CFR 60-2 (revised order no. 4).
3. A. J. Jaffe, R. M. Cullen, M. G. Powers, J. C. Ridley, H. F. Spirer, G. Stolnitz, and J. A. Turner, Summary notes on the statistics of federal affirmative action programs, *American Journal of Economics and Sociology, 41*(4):321 (1982).

4. U.S. Bureau of the Census, *1980 Census of the Population, Characteristics of the Population, General Social and Economic Characteristics U.S. Summary,* Part 1, PC 80-1-C1, 1983, Table 74.
5. U.S. Bureau of the Census, *1980 Census of the Population, Characteristics of the Population, General Social and Economic Characteristics, U.S. Summary,* Vol. 1, Chapter C, Part 1, PC 80-1-C1, 1983.
6. Charles R. Mann, "Abuses of statistics in civil rights legislation," 1980 Annual Meeting, The American Statistical Association, Houston, 1980.
7. Wilfrid J. Dixon and Frank J. Massey, Jr., *Introduction to Statistical Analysis,* 3d ed., McGraw-Hill, New York, 1969.
8. Robert Hooke, *How to Tell the Liars from the Statisticians,* Marcel Dekker, New York, 1983.

CHAPTER 13: ECTOPLASTISTICS

1. Donald J. Treiman and Heidi I. Hartmann (eds.), *Women, Work and Wages,* National Academy Press, New York, 1981.
2. Fernand Braudel, *Civilization and Capitalism, 15th–18th Centuries, Volume III: The perspective of the world,* Harper & Row, New York, 1984.
3. U.S. Department of Labor, Bulletin no. 40, May, 1902.
4. Joseph P. Goldberg and William T. Moye, *The First Hundred Years of the Bureau of Labor Statistics,* U.S. Government Printing Office, 1985.
5. U.S. Department of Labor, Bulletin no. 3, March, 1896.
6. U.S. Bureau of the Census, *1980 Census,* Vol. PC 80-1-D1-A, 1980.
7. Setback for the doctrine of pay equity, *New York Times,* June 23, 1985.
8. Women in state jobs gain in pay equity, *New York Times,* May 20, 1985.
9. Los Angeles backing equal pay for jobs of "comparable worth," *New York Times,* May 9, 1985.
10. Max Singer, The vitality of mythical numbers, *The Public Interest, 23*:3 (1971).
11. Marc Galantner, Reading the landscape of disputes, *U.C.L.A. Law Review, 31*:4 (1983).
12. U.S. will follow legal precedent in benefit cases, *New York Times,* June 4, 1985.
13. Jeffrey Schmalz, U.S. told to use court precedent in denying aid, *New York Times,* Aug. 20, 1985.
14. William Abraham, Letter from Ethiopia, *New York Statistician, 27*(3):3 (1976).
15. Office of the Population and Housing Census Commission, *Ethiopia, 1984, Population and Housing, Census Preliminary Report,* Vol. 1, no. 1, Addis Ababa, 1984.
16. Paul Brodeur, Annals of law, the asbestos industry on trial, Parts I to IV, *New Yorker,* June 10, 17, 24, and July 1, 1985.
17. Andrew Pollack, Electronic trespassers, *New York Times,* Nov. 10, 1983.
18. William Greider, The education of David Stockman, *Atlantic Monthly,* Dec. 1981.

CHAPTER 14: BIG BROTHER/SISTER

1. Garcilaso de la Vega, El Inca, *Royal Commentaries of the Incas, Part One,* Harold V. Livermore (transl.), University of Texas Press, Austin, 1966.

2. C. Taeuber, "Census," *International Encyclopedia of the Social Sciences,* Vol. 2, Crowell, Collier, and MacMillan, New York, 1968.
3. Martin Tolchin, Pick a number, any politically powerful number, *New York Times,* June 5, 1984.
4. *SCOPE,* Creative Journals, Ltd., London, October, 1944.
5. M. C. Buer, *Health, Wealth, and Population in the Early Days of the Industrial Revolution,* George Routledge & Sons, London, 1926.
6. *New York Times,* Aug. 25, 1985.
7. Robert A. Lewis, The universality of demographic processes in the USSR, George J. Demko and Roland J. Fuchs (eds.), *Geographical Studies on the Soviet Union, Essays in honor of Chauncy D. Harris,* University of Chicago Press, Chicago, 1984.
8. Fred W. Grupp and Ellen Jones, Infant mortality trends in the Soviet Union, unpublished manuscript, 1983.
9. Clifford D. May, An ancient disease advances, *New York Times,* Aug. 18, 1985.
10. A. J. Jaffe, Analysis of international statistics, *New York Statistician, 27*(3):5 (1976).
11. Congressional Research Service of the Library of Congress, for the Committee on Government Operations, *An Update on the Status of Major Federal Statistical Agencies, Fiscal Year 1986,* U.S. Government Printing Office, 1985.
12. Congressional Research Service of the Library of Congress, *The Federal Statistical System 1980 to 1985,* U.S. Government Printing Office, 1984.
13. *Hearing before the Subcommittee of Committee on Census and Population of the Committee on Post Office and Civil Service,* House of Representatives, 97th Congress, second session, March 16, 1982, Serial no. 97-41.
14. Martin Tolchin, U.S. plans cut in data collection and distribution, *New York Times,* March 31, 1985.
15. Robert J. Wolfson, Reagan administration attacks the facts, *New York Times,* Apr. 11, 1985.
16. Martin Tolchin, Librarians rally the troops to battle U.S. on information, *New York Times,* Apr. 8, 1985.
17. Lazare Teper, Politicization and statistics, *New York Statistician, 25*(4):1 (1974).
18. U.S. Bureau of the Census, *Statistical Abstract of the United States: 1972.*
19. A. J. Jaffe, Politicians at play, *New York Statistician, 34*(4):1 (1983).
20. Richard Reeves, Journey to Pakistan, *New Yorker,* Oct. 1, 1984.
21. United Nations, *United Nations Year Book of National Accounts,* Vol. III, New York, 1972.
22. United Nations, *United Nations Year Book of National Accounts,* Vol. III, New York, 1981.
23. Chen Ta, *Population in Modern China,* University of Chicago Press, Chicago, 1946.
24. Leo A. Orleans, Chinese Statistics: the impossible dream, *The American Statistician, 28*(2):47 (1974).
25. Press in China admits to lies, boasts and puffery, *New York Times,* Aug. 29, 1979.
26. Fox Butterfield, *China,* Bantam Books, New York, 1983.
27. Christopher S. Wren, Chinese press tries to mend soiled image, *New York Times,* May 20, 1984.

28. John F. Burns, A new team checks in at the Great Wall Hotel, *New York Times,* March 24, 1985.
29. China says economy soared in 1985, *New York Times,* March 10, 1985.

APPENDIX: LIFE TABLES

1. U.S. Bureau of the Census, *Statistical Abstract of the United States: 1985.*

Index